SIMPLE LESSONS

FOR

THE USE OF TEACHERS

IN

INFANT SUNDAY SCHOOLS.

FOLLOWING THE CHURCH SEASONS
ADVENT TO TRINITY.

BY

EDITH EMILY BAKER,

SUPERINTENDENT OF THE INFANT SUNDAY SCHOOL
OF ST. GEORGE, BLOOMSBURY.

PUBLISHED UNDER THE DIRECTION OF THE TRACT COMMITTEE.

LONDON:
SOCIETY FOR PROMOTING CHRISTIAN KNOWLEDGE,
NORTHUMBERLAND AVENUE, CHARING CROSS, W.C.;
48, QUEEN VICTORIA STREET, E.C.;
26, ST. GEORGE'S PLACE, HYDE PARK CORNER, S.W.
BRIGHTON: 135, NORTH STREET.
NEW YORK: E. & J. B. YOUNG & CO.
1885.

PREFACE.

These lessons, written in the prayerful trust that God's blessing may be upon them, are intended to suit a school of little children from four to seven years of age, divided into classes all learning the same lesson; simultaneous repetition of the text to the Superintendent, with short questioning upon it, during a ten-minutes' break in the teaching, is a part of the plan.

3, MONTAGUE PLACE, W. C.

CONTENTS.

viii   CONTENTS.

Advent Sunday.

Verse to learn—

"Therefore, be ye also ready; for in such an hour as ye think not the Son of Man cometh."—*St. Matt.* xxiv. 44.

Key-note *ready*. "Son of man"—Son of God too, Jesus Christ. Where is Jesus now? Disciples saw Him go away (Acts i. 9). And one day He will come again (verse 11). Sure to come, for He said so Himself (St. Matt. xxiv. 30). Do not know when; but even if we are dead and asleep, we shall wake to see Jesus when He comes.

Suppose kind friend, Aunt or Teacher, told children she would come and take them to nice happy place (describe and enlarge upon pleasures of going), what would children do? (1) Get ready; (2) Keep ready.

(1) Get ready—wash, put on best clothes, mended, clean. (2) Keep ready—remember, wait patiently, watch.

This only story. Jesus coming, real, true: better than Aunt or Teacher. You must be ready; have your sins washed away, your souls made clean and white. When you do bad, naughty things tell Jesus about it; ask Him to forgive you, and help you.

If Teacher came all of a sudden, after waiting a little, would she find you had forgotten all about it, were quarrelling, were cross, dirty, not fit to be seen? No; I

think you would watch and keep ready—smiling face, run
to meet her.   So try to keep ready for Jesus; remember
His coming, think about it; you might look up into sky
every day, and say softly "Jesus is coming, I must be
ready."   Try every day to be good, because He loves to
have you good; then you will be so glad when Jesus
comes—you will be *ready*.

(If lesson wanted longer, illustrate by parable of Wise
and Foolish Virgins, St. Matt. xxv; or the Good and Bad
Servants, St. Matt. xxiv.)

Second Sunday in Advent.

THE BIBLE.

" And that from a child thou hast known the Holy Scriptures, which
are able to make thee wise unto salvation, through faith in Christ
Jesus."

There was once a little boy, Timothy—had good mother
and good grandmother (enlarge); they taught him—what?
Perhaps many things; but one we know, and it was the
best thing—taught him to know the Bible.   What is it
called in verse —" holy "—? Scripture means *writing*—
holy writings.   Why "holy"?   Book of God, *God's
Word.*   What use was it teaching Timothy to know his
Bible?   It could make him " *wise* unto salvation through
Jesus Christ." (Recapitulate and say verse.) Now are *we*
learning to know our Bible?   Thank God every little

child here may learn to know the Bible, if it will only
listen.   Once very few had Bibles or could read—big
Bible chained to Church desk—people walked miles to
read one text.   No little Bibles—cost a great deal of
money.   Now, even you little children have got Bibles.
I have one: you, and you, have one; or will have one
by and by.   Whose Word is it?   Then do not treat it
like a common book—*very* carefully, *very* gently; try
not to let it get torn or dirty; treat it as if it were gold,
more precious.   What did Bible do for Timothy?—
wise through—?   Jesus Christ.   Bible tells us about Jesus
Christ, His life on earth: how He died for us, rose again;
tells how we may try to please Him.   We must listen to it,
learn it, read, attend to it, ask God to let it make us
" wise unto salvation through Jesus Christ."

Third Sunday in Advent.

REPENTANCE.

Verse to learn—

"The word of God came unto John the son of Zacharias in the
wilderness. And he came into all the country about Jordan, preach-
ing the baptism of repentance for the remission of sins."—*St. Luke*
iii. 2, 3.

Heard about wonderful *Book* last Sunday—Bible.
Now hear about wonderful *man*—John the Baptist.   God

speaks by His Book, so Bible sometimes called the Word of God. But God speaks sometimes in other ways, so in the text we are told "the Word of God came to"--? (John). He was to be God's Messenger (Mark i. 2).

1. Who was John?

2. What did he do?

1. John had a wonderful history. God had thought about him, and prepared him (made him *ready*) for what he had to do, from the very first, when he was a tiny baby. Good father and mother, doing right, thinking of what God wanted them to do; not their own way, but God's way. What does Catechism say about our baptism promise? (3rd) "Keep God's holy will and commandments, and walk in the same all the days of my life." That is what John's father and mother did (St. Luke i. 6); they were old and had gone on doing well all the days of their life. Father's name? (see verse). Mother's name? (Elisabeth). Angel came; told them about the baby, what his name was to be. What was it? How did the Angel know? Angel a heavenly messenger sent by God, who was preparing that little baby John to be His messenger on earth. And by and by John grew up to be a man, strong, hardy, not caring for grand soft clothes to wear, or nice things to eat, thinking only of God's work he was to do, the message he was to take. We see what John *was*; now—

2. What did he do? (repeat text). Now the time was come; he could not speak till God gave him the

message, for he was God's messenger. What was the message?—" the baptism of repentance for the remission of sins." To use one word, he was to call upon all the people to *repent*? To—?

To repent of sin means to be sorry enough to make us leave it; to give it up, have no more to do with it. John called "the Baptist" from what our text says. (Repeat.) Who was coming after John, do you think? Christmas coming soon to tell us of His coming, greater than John; coming with message of love from God (1 John iii. 2); to prove this love by dying for us. But before we can take the message of Jesus into our hearts, we must remember John the Baptist's message. Sin such a bad hateful thing in God's sight. (Teacher may enlarge as to special sins of children.) Ask God to help us to repent; to be sorry enough to give up our sins, whatever they may be.

Fourth Sunday in Advent.

WHY JESUS CAME.

Texts to learn—

"God is love."—1 *St. John* iv. 8.
"We love Him because He first loved us."—1 *St. John* iv. 19.

(Babies' class to learn first verse only. Every child say it to Teacher by itself.)

In a few more days Christmas will be here, the happy

day we keep as the Lord Jesus Christ's birthday, when
we specially remember how He came into this world as a
little child.

"There came a little child to earth,
Long ago.
And the Angels of God proclaimed His birth,
High and low.
Out in the night so calm and still
Their song was heard,
For they knew that the Child on Bethlehem's hill
Was Christ the Lord."

To-day we are going to think *why* He came. Why
did God the Father send His dear Son into the world to
be a poor man, to suffer, and to die, for us? It was because
of His *love* for us—God is love. God made a beautiful
garden for Adam and Eve; made them happy. Did
they keep happy? Disobedient! God had to punish
them; but still he loved them, and promised to send
some One to be their Saviour. When we are naughty,
we have to be punished, and God is sorry for us. God
is holy; cannot bear bad things. No sin can go to
heaven; but He loves us still, and will let Jesus take
away our sins, if we ask Him. What did God promise to poor Adam and Eve? A Saviour. People
waited long, lived, died—Noah, Abraham, Moses, Israelites, David, the Prophets—God did not forget, *never*
forgets; helped them, taught them, loved them: and
at last the Saviour came; and now everybody may
know about God's great love to men.

(Repeat second text.)

When we see a kind face smiling at us, what do we do? smile back. When kind friends—father, mother, teacher—love us, we—? love them back. So when we think of God's love to us, we lift up our hearts to Him, and "love Him because He first loved us."

Christmas Day.

GLORY TO GOD.

"Glory to God in the highest, and on earth peace, good will toward men."—*St. Luke* ii. 14.

(The time on Christmas Day is usually short, and partly taken up in mutual greetings and natural overflow of happiness on the part of the children; the Christmas story is generally well known, but it might be well to draw it out by a few questions as to *why* we are so happy, and *whose* birthday it is, before going on to the following lesson.)

Ever seen very beautiful church? Westminster Abbey, St. Paul's, York Minster? Beautiful service? (Describe Sons of Clergy Festival, or any great service you may have heard), sweet singing, praise to God, very beautiful to hear; but what these poor shepherds saw and heard was far more beautiful. (Describe Eastern shepherds guarding flocks from harm, sitting up at night, out in the fields.) Angel came—could see him; perhaps Angels come near us sometimes, but we cannot see them.

Shepherds saw them, all bright and glorious (verse 9), and heard good news—*Saviour* come, the promised Saviour—"And suddenly there was with the Angel a multitude of the heavenly host praising God and saying—"(repeat text). Angels do God's will. What do we say in the Lord's Prayer? "Thy will be done on earth, *as it is in heaven.*" We long to do God's will as the Angels do it. The Angels love to praise God, and to bring good news like this to men. Glorious and happy thing to praise God; and in his love and mercy He *likes* to hear His children praise Him, and say as the Angels did, "Glory to God in the highest, on earth peace, good will toward men."

(If time allows, children might repeat hymn "While shepherds watched their flocks by night.")

Sunday after Christmas.

THE NAME OF JESUS.

Verse to learn—

"And thou shalt call His Name Jesus, for He shall save His people from their sins."—*St. Matt.* i. 21.

How many days since Christmas? (Count on till same day of the week is reached)—now how many? Eight.

What was done to little Jewish boys when eight days old? Circumcised, and had name given them. And the Lord Christ was circumcised and had a name given Him when He was eight days old, because He was born a little Jewish baby. What was the name given to Him then? Angel told His Name. What did Angel say? (repeat verse). Wonderful, precious Name! some day perhaps you will learn beautiful hymn that begins "How sweet the Name of Jesus sounds." Jesus= Saviour; to save His people from—? their sins.

And as we love the Name of Jesus, so we must reverence it too; never say it lightly, carelessly. Where are we told not to take God's Name in vain? (Repeat Third Commandment.) Jesus is God; we must love, honour, and keep holy His Name. Where do we pray that God's Name may be hallowed=kept holy?

Christian people's babies not circumcised now; but we had names given--when? Baptism. Jesus said the children might come to Him (St. Matt. xix. 14), so they are brought to be baptized. God's grace given to them in answer to prayer, and Christian name. (See Baptismal Service, and how child called at once by new Name —"John, I baptize thee.") Before baptism known as Baby—neighbour comes in, says "How is the baby?" Now it is Mary, Jane, Harry. Each one has own particular name. We know our own name, don't we? never forget it! *Jesus knows it too* (St. John x. 3).

Three things to remember—

1. The text about Name of Jesus.

2. Our names given in Baptism.

3. Try never to do anything to make us ashamed of our names, for *Jesus knows them.*

First Sunday after Epiphany.

NUNC DIMITTIS. I.

" Lord, now lettest thou thy servant depart in peace, according to Thy word :
For mine eyes have seen Thy salvation,
Which thou hast prepared before the face of all people ;
A light to lighten the Gentiles, and the glory of Thy people Israel."
—*St. Luke* ii. 29-32.

Have you ever seen little baby in church, brought by father, mother, and friends ? What has it come there for ? To be baptized, and received into Christ's Church. Jews had a law that little Jewish boys should be brought to the Temple and presented to God. What wonderful Babe was it that was born on Christmas Day, became a little Jewish boy ?—Jesus. Do you think His mother did what other Jewish mothers did ? Yes—brought Him to the Temple—God's House. Found an old man there, Simeon ; good old man, took the Babe in his arms, and blessed God. Who was that Babe ?—Jesus. Simeon knew that ; that was why He thanked God, and said this beautiful Psalm. (Repeat.) Teacher must paraphrase verses simply. Simeon spoke to God, said he could die happily, and in peace, for he had seen Jesus as God had promised him; his eyes had seen—? Jesus—salvation— " save His people from their sins."

Where is Jesus' now? Not little Baby now, great King. We cannot see Jesus, as Simeon did; but Jesus is our salvation, and we can believe as Simeon did, and thank God in the very words Simeon used.

Three things to remember :—

1. Simeon in the Temple—nice thing when old men go to church. (Psalm cxlviii. 12.)

2. Found Jesus there · Jesus in church with us though we cannot see Him.

3. Believed in Jesus and thanked God.

Second Sunday after Epiphany.

NUNC DIMITTIS. II.

Lesson upon verse 4, "A Light to lighten the Gentiles."

For very young children, talk about light—pretty star in the East—wise men wanted to see the dear Baby. God showed them the way; made pretty star for them.

Makes good things for us; helps us to think about Jesus, to find Him, and love Him.

For the older classes describe long journey of the Magi—difficulties in finding the way. Whom did they seek? *Jews* knew where He should be born; these men not Jews, Gentiles. Are we Jews? Most of us in this country are Gentiles too. God helped them to find Jesus by a star. Will help us to find Jesus.

1. In the Bible, read about Him. Listen when Bible read in church, and when Teacher tells about Jesus.

2. In prayer. Where is Jesus now? In Heaven. Same Jesus, not Babe of Bethlehem now, but Lord in Heaven; let our prayers go up and meet Him there.

Third Sunday after Epiphany.

JESUS SHOWN BY HIS WORKS.

"Now when the sun was setting, all they that had any sick with divers diseases brought them unto Him; and He laid His hands on every one of them, and healed them."—*St. Luke* iv. 40.

Heard last Sunday how Jesus was shown by a star; now hear how Jesus was shown by His *works* (Gospel). Can any of you remember the Messenger of God who came before Jesus? (See Lesson, Third Sunday in Advent.) What was his name? He once sent two of his friends—

disciples—to ask Jesus if He were really Christ the Son of God, and Jesus told them to tell John what they had *seen*. Now what was it they saw, things mentioned in our text, and more too—the blind made to see, lame to walk, sick made well, deaf to hear, dead to live, poor had good news told them, all got just what they most wanted. (Enlarge and bring home to children by illustration the misery of the affliction and joy of deliverance.) Now when the two saw this, and told their Master John, what would he be sure of?—that Jesus was the Son of God.

1. Because only God could do things like this.

2. Because the Bible said Christ would do these wonderful things when He came (Isaiah xxxv. 5); and Jesus did, you see; so John the Baptist and everybody could be very sure. The works He did showed about Him (John xiv. 11).

Picture scene in text—sun setting, people hurrying to come before it was too—? (dark). Hurrying to do what? Bring sick to Jesus. Three things to notice :—

1. Brought them unto *Him*. Who? Jesus: no one else would do—Jesus only—no other Saviour.

2. *Divers* diseases—different bad things, not merely one sort, all sorts. God is almighty, able to do—? (everything). Jesus is God, so is almighty, can do everything, heal everybody, *all* trouble.

3. Laid His hands on every *one* and healed them. Not one sent away—none too little or too bad.

These people wanted help. Got it from Jesus.

We want help. Sin our disease—very bad, must die if not healed. Can get help from Jesus.

1. Jesus only. Pray to Him; nobody else to cure sin.

2. Divers diseases. Every kind—lies, temper, disobedience.

3. Every one. None too small, none too bad —every one.

Fourth Sunday after Epiphany.

JESUS SHOWN BY HIS POWER OVER NATURE. STILLING THE TEMPEST.

"What manner of man is this, that even the winds and the sea obey him !"—*St. Matt.* viii. 27.

Give instances of *God's* power over nature—*whale* for Jonah; *ravens* commanded to feed Elijah (1 Kings xvii. 4, 6); the *lions* which were not to eat Daniel (Dan. vi. 22); *wind* to blow away plague of locusts (Exod. x. 19); *sea*—Red sea going back (Exod. xiv. 21); *river*—Jordan standing up like a wall (Josh. iii. 16). And children themselves ought to be able to give *star* which showed the way to Jesus. All these obeyed God: He made them and has power over them, and they do what He tells them. Ever seen the sea ?—much bigger than water in the Park ; waves washing up along the side, getting big and rough when wind blows, and making little boats and

even big ships toss about; sometimes so strong it dashes
them to pieces, and the poor people are drowned.   Once
Persian king angry with the sea, told his servants to beat
it.   Did it mind him?   No; he not master of the sea.
Who is?   Picture scene on Sea of Galilee—little ship,
disciples, Jesus asleep.   Great storm, disciples very frightened; knew *they* could not make sea quiet; did not
believe Jesus could, but woke Him, " Lord, save us, we
perish!" we are going to die! How?   Be drowned.
Jesus heard (does He always hear when people cry to
Him? now?); spoke back to them, asked why they were
frightened when He was there; got up and told the
winds and the waves not to be so rough, and they were
quiet directly.   Then the men wondered and said—(repeat text).   *We* can answer: Jesus is *God*, that is why
the winds and waves obey Him.

Side lesson if desired.   Jesus not pleased with them for
being frightened when He was there to take care of them;
not pleased if we are frightened, for He is ready and able
to help us.   Some people frightened in thunderstorms,
on being left alone, or in the dark.   Do not let us be so—
Jesus is near us; He knows all about storms; He can
see us through the darkness; He will do the very best
that can be done for us, for He loves and cares for us;
and He is God whom all things obey.

Fifth Sunday after Epiphany.

JESUS SHOWN BY HIS WORDS.

"And Jesus came down to Capernaum, a city of Galilee, and taught them on the sabbath days. And they were astonished at His doctrine: for His word was with power."—*St. Luke* iv. 31, 32.

What a wonderful Sunday school! Jesus the Teacher! We have seen how Jesus was shown by His *works*, healing the—? by his power over nature—making the sea quiet, stilling the tempest; now by His teaching. What does our text say the people felt when they heard Jesus? "Astonished"—His words were full of might and truth, He "spake as never man spake." Why? *God* as well as man. Would you like to have been in Jesus's Sunday school? He told the people beautiful stories sometimes, each one to teach them something. You shall hear one to-day. Any of you been in the country?—seen fields with things growing, turnips? grass? *corn*? Jesus lived in the country—often saw corn growing. What is a man who grows corn called? Once there was a man who had fields, and wanted to grow something good in his fields: what must he do to get anything to grow? Put seed in. He wanted corn, so put corn seed in; good seed, so he hoped to get—? good corn. If teacher can get ear of corn to show seed, also weed to explain tares, so much the better. Tell how enemy came in the night—dark, every one asleep, sowed bad seed of weeds, tares (bearded darnel, the only grass with poisonous seed entirely like wheat till ear comes). Bad, cruel thing to do! Wheat grew from the good seed, but what from the bad seed? grew together —just alike *out*side, but very different *in*side; one good, the other—? one to be gathered in to make a beautiful harvest, the other to be taken up and burnt. This not quite all the story, but enough for our lesson to-day.

We may be like the good wheat or the bad tares— bad tares if we let Satan come, as he did to Eve, and sow bad thoughts in our hearts; good wheat if we ask God to put good seed in.

Sixth Sunday after Epiphany.

WHY CHRIST WAS MANIFESTED.

"For this purpose the Son of God was manifested, that He might destroy the works of the devil."—*St. John* iii. 8.

Sixth Sunday after—? Epiphany. Look at your Prayer-book, find the Epiphany; read what is written underneath to explain. The Epiphany or—? Manifest-ation of Christ to the Gentiles. " Manifestation " long word too; means showing forth, just what we have been learning about—Jesus shown by a star—by His works— by His power over nature—by His words. (Recapitulate previous lessons.)

And now *why* was it? Our text tells us—"for this purpose"—this reason—"to destroy the works of the devil"—that is, sin—what you promised to renounce when you were baptized. (Repeat "renounce the devil and all his works," etc.) Cannot do this of ourselves, so Jesus was manifested, not only to renounce but to destroy. Suppose you were a prisoner in giant's castle, chained up, high walls, cruel things going on all round, misery and wretchedness; great prince came, bound the wicked giant, destroyed his works, pulled down the walls, set you free—how happy, how grateful you would feel! This is true: we were prisoners of the devil "tied and bound with the chain of our sins," high wall made by sin between us and God—Jesus came, Prince of Peace, set us free, destroyed the works of the devil. Should we not love and thank Him? and try to do His will?

Septuagesima Sunday.

THE RACE.

"So run that ye may obtain."—1 *Cor.* ix. 25.

Ever heard about a race, or seen one? Oxford and Cambridge boat-race, when you see the cab and omnibus drivers with bits of light or dark blue ribbon, to show which they wish to win? Or the boys running at the summer school-treat? You know about that, how they

stand in a row, all ready, listening—one, two, three, off!
Away they go, running as hard as they can. What are
they trying to do? To get to some particular place first;
to get or *obtain* a prize. Who is first; that one; no, see
he has tumbled down; now two close together. On they
go; one makes a rush, is in, has obtained the prize!

How did he run? *Well*—good start, good run, good
finish.

1. Good start—set off at once, when he heard the bell
or signal; did not say, "Oh, wait a bit, I shall have
plenty of time by and by."

2. Good run—steady all the way; if he fell down got
up again directly; did not forget what he was about,
or run after other things.

3. Good finish—joyful time, hard work of running
over, prize is his, rest may follow!

Now, what is our text? St. Paul speaks—who to?
Not boys running race for play, but Christian people at
Corinth; puts it in his letter to them—good advice;
and it is read to us in church to-day (see Epistle) as
good advice for us too.

Our Christian life like a race, begun at—? Baptism.
Ended—? When our life on earth ends. Who gives
the prize? What is it? Rev. ii. 10, "*I* will give thee—
crown of life." Is it worth having? Yes; it is a
"crown of glory that fadeth not away." Would you
like to get it? Then if you want to get the crown *then*,
you must "so run" *now*.

1. Ask God to help and make you strong, keep you from falling.

2. Go straight on doing right, no matter whether hard or easy.

3. Give up anything that keeps you back—bad habits or bad companions.

Sexagesima Sunday.

THE PARABLE OF THE SOWER.

"Now the parable is this: The seed is the Word of God."—*St. Luke* viii. 11 (Epistle for the day).

Who was the most wonderful Teacher? Remember lesson about Jesus's wonderful words (Lesson ii), showed that He was God as well as man. When did He teach? Where? Should you like to have teaching from Jesus? He will teach you, if you ask Him—two ways :—

1. Give you His Holy Spirit in your heart—hear more about that on Whitsun-Day.

2. By His Word—read in Bible, or taught you by your teachers.

One of Jesus's beautiful stories about this very thing. Remember the one about two kinds of seed? This about one kind of seed, different kinds of ground. (For London children the Square garden or Park must be used as illustration—seed falling on path, stone curb of railing, and prepared flower-bed; to village children the field, the wayside and rocky place will all be familiar.)

Paraphrase story simply—fowls of the air, birds, so fond of following and pecking up any uncovered seeds; lacked moisture, no nice damp earth for its roots; hundred-fold one little corn seed, full wheat ear; story with a meaning—parable.

Repeat text. Seed is what? Sower whoever gives us the seed—perhaps clergyman in sermon, perhaps your teacher *now*. What do you hear the words with? What do you think of them with? Your heart: that is the ground where seed goes. Is it good ground?

Verses 12–15. Very different sorts of ground. First three, seed wasted, comes to no good; sower disappointed, ground bare, no lovely fruit to spring up. Very sad! Worse when it is God's Word. Even in this schoolroom there may be the different sorts—some ears only, heart does not listen at all. Some hear—go away and forget; and others think of other things, choke the good seed. Are you like that? will you waste the good seed? Ask God to make your heart good ground, that the word may spring up, and grow, and bring fruit a hundred-fold.

Quinquagesima Sunday.

JESUS SHOWN BY HIS POWER IN OUR HEARTS.

"And now abideth faith, hope, charity, these three; but the greatest of these is charity."—I *Cor.* xiii. 13.

Have you ever seen a beautiful tree covered with fruit? like the nice trees in the Square gardens, only rosy apples, or pears, or pretty cherries on it? You remember God made beautiful fruit-trees in the garden where Adam and Eve were (Eden), Gen. ii. 9. (Read that verse.) "Every tree pleasant to sight and good for food." Pleasant to sight, like—? Lilac, Laburnum, May. Food? fruit-trees, palm-trees, cocoanut, bread-fruit.

Look further—two specially mentioned, tree of life, and tree of knowledge of good and evil. Should you like to see the beautiful tree of life? It is growing in Heaven Rev. xxii. 2, beside the beautiful river. Who will see it and eat its good fruit? Those that overcome (Rev. ii. 7); those that win in that race we talked about two Sundays ago. Are you running well?

What have we been talking about to-day? Trees—Tree of Life. Now going to speak of something that is *like* a tree—root, leaves, fruit (repeat text—faith, hope, charity).

Faith, like the root, goes down deep and takes hold, keeps the tree steady and feeds it; cannot live without root, so we cannot be good and please God without faith. (Heb. xi. 6.)

Hope is like the beautiful green leaves; they make a pleasant song with their rustle in the wind, and let the sunshine dance in through their cool shade: they show the tree is alive, and make people look forward to the fruit.

What is the fruit?—Love=charity. Charity-school where parents do not pay—children taught for *love*. Sunday schools are all love schools. Sweet, pleasant love, always thinking kind things, saying kind things, doing kind things! let us try to be like trees with plenty of pleasant fruit of love. How? by trusting in God, and looking to Him to help us (see Jer. xvii. 7, 8: says it is like tree planted by river)—*roots* spreading out, *leaf* green, no end to the *fruit*.

First Sunday in Lent.

SIN. ADAM'S SIN.

"Wherefore as by one man sin entered into the world, and death by sin, and so death passed upon all men, for that all have sinned."— *Romans* v. 12.

Any of you know what last Wednesday is called? Ash Wednesday—first day of Lent (see Collect). This the first Sunday in Lent—special time to think of our sins, to do what John the Baptist told the people, to—?

repent. Sin is so bad, so dreadful, must think specially
sometimes, as well as being sorry at once when we do
wrong. Think what is told us about sin in the Bible.
(Repeat text.) Sin *entered* into the world, came in;
then once it was outside—away—not in the world at all:
oh, happy world when no sin was in it, all good (Gen. i. 31),
all lovely and delightful! Who brought sin in? "one
man"—Adam. Most children familiar with story of the .
Fall; draw it out by questions—the garden: remind them
of the beautiful trees spoken of last Sunday—the special
ones, the one they must not touch, the disobedience, the
shame, the punishment, present thorns and thistles; toil,
care, pain for body; future—death for body. Very sad
for Adam and his wife Eve; sad to lose their happy
garden, to see the beautiful tree of life guarded from
them by Angel with flaming torch: but that was not the
worst—the *sin* in their hearts was the really bad thing;
it made them even want to hide from God, the good God
who had made all these beautiful things for them, the
holy God Who cannot bear sin. And the next bad thing
was that this terrible sin was like some bad catching
disease; Adam's son had it—how did he show it? other
children had it, and their children down to our very own
selves. Adam let sin come into the world, and it *stayed
in*; no country without it, no town, no house, no heart—
in this room, children irreverent at prayers, not obedient
to Teachers, not kind, not listening to hear about God,
shows sin is here—very sad thing. How did sin get

into world? Did it make Adam happy? Cain happy? *thought* they should be; deceived by Satan who loves sin. God hates it, for God is good. Ask God to make you hate sin (Ps. xcvii. 10). Ask Him to show you how Jesus will set you free. Ask him to give you His Holy Spirit in your hearts to make you good.

Second Sunday in Lent.

SIN. WORLD'S SIN IN NOAH'S TIME.

"And God saw that the wickedness of man was great in the earth."
—*Gen.* vi. 5.

> "'Twas but one little drop of sin
> We saw this morning enter in,
> And lo! at eventide the world is drowned."

A rush, a roar, a terrible flood of water carrying all before it, sweeping away houses, bridges, men, women, children—what is it? reservoir burst, place where water stored up in immense quantity. All safe a while ago, happy peaceful homes; and now destruction, fear, and death—how was it? what began the mischief? A tiny crack, two or three drops oozing through, little hole that one of your small fingers might have stopped; tiny stream flowing faster, faster; bigger, bigger grows the hole, crack go the sides, away goes the mighty flood, destroying all in its way. Perhaps there are people in little

D

cottage near—oh, they will be drowned! No, see, there is a boat; it comes near, they get into it; they are safe! *Sin* like this; began with one man. Who? It grew and spread till its misery drowned all the earth. (Repeat text.) *Wickedness*, what is the short word for it? *sin*. Great sin; and God saw it. God must see it, God sees—? every thing. Did God see Adam's sin? Yes; and he saw too that Adam was afraid and ashamed; had mercy on Adam and Eve and gave them promise as well as punishment. Wickedness of *these* men (text), so dreadful, no fear, no shame that God should see. Ask God that you may always be ashamed when you have done wrong; much better to go at once and say you are sorry, and be forgiven, and try hard to do better. There was one man who did try, and God saw him too, in the midst of all those wicked men. God saw that one good one, and saved him in a very wonderful way. (Question out story of Noah, generally familiar to children.) Why was every-one who was not in the Ark drowned? Who saw their sin? Who hates it? Who loves sin and tries to tempt people to follow him? Oh how earnestly we should pray that we may hate sin and follow Christ—remember God always sees sin, "Be sure your sin will find you out." What saved Noah? Ark; got inside and was safe. Have we any hope of being safe?—"safe in the arms of Jesus." What makes us want a Saviour? what did He come to save his people from? Let us hate our sins and love our Saviour.

Third Sunday in Lent.

SIN. ISRAELITES' SIN.

"But for all this they sinned yet more, and believed not His wondrous works."—*Psalm* viii. 32 (Prayer-book version).

What terrible bad thing have we been talking of last two Sundays? Who first let sin come in? how? Who else had it in their hearts? Did it get better or worse? Who had it in Noah's time? Who saw it? How did God punish the wicked? Very sad, but no punishment too bad for sin. What is sin? disobedience to God's laws, not doing as He tells us, breaking His commandments. Had Adam a command? What? Did He break it?—this was *sin*. Has God given you children any commands? Obey your parents, Love one another, Love God, Honour His Name, and many more. When you disobey these, even a little, you *sin*, the same sad sin that turned Adam and Eve out of the garden and caused the Flood. God promised there should never be any more flood like that; more people came, grew up, more and more, till the earth was full again, some good, some bad. God let them grow together, like wheat and tares (see Lesson 11).

At last God chose out a good man. Can God see when any one tries to be good? Remember how He saw—Noah. Chose out Abraham, gave him a son Isaac, and grandson Jacob; and from them made a people,

Israel—God's chosen people. He loved them and gave them great promises for the future, many good things for the present; one promise was to give them a beautiful country, Land of Canaan, the Promised Land. And God Himself said He would take them there. Now how did they behave? did they love God for His goodness to them? Not always. Give instances, God's goodness, Ps. lxxxviii. 14–17; then comes verse 18, "yet they sinned, even grumbled against God their Friend, and turned away to false gods, idols of wood and stone." Then they were sorry, and God did more good things (verse 28, 30); but for all this (verse 32), sinned yet more! What do we call it when any one is not good to those who are kind to them? Un—grateful. God very good to His people Israel; they ungrateful—how very bad! Is God good to us? How? (Tell children to mention blessings.) Then if we sin against God we are ungrateful too! How we ought to hate sin, if it make us ungrateful to our good God!

Fourth Sunday in Lent.

SIN. OUR SIN.

"I will arise and go to my Father, and will say unto Him, Father. I have sinned against Heaven, and before Thee, and am no more worthy to be called Thy son."—*St. Luke* xv. 18, 19 (and in Prayer-book sentences).

These words first said in one of the beautiful stories of Jesus. Prodigal Son. Describe—wealth, going away,

waste, want, repentance, return; said "I will go," and he
did go. How did he feel? ashamed and sorry, no excuse, no
untruths—"Father, I have sinned." *Who* said this? Prodigal Son in the story; yes, but turn to Prayer-books—
yours, and yours—each one their own, and see something
said by Minister, to be attended to by each one of us,
"Father, I have sinned"—sinned against our Father in
Heaven; and when we come to Him we say humbly and
sorrowfully, "Father, I have sinned." Who *first* sinned?
second? We may begin long list with Adam, Cain, and
the Israelites; but we must come at last to "Father,
I have sinned."

Ever seen flock of sheep being driven? Some years
ago they used to come through the streets of London in
the daytime, now only very early in the morning. When
I was a little girl, out with my nurse, often saw sheep
driven along the road in amongst the carts and cabs, and
sometimes they would give the man (the—? shepherd)
great trouble by straying away, running up some side
street where they were not meant to go, or even into a
shop; they do the same sometimes in the country fields,
stray away, get through the gate or hedge away from
their shepherd. Poor silly sheep! they can only be safe
with their shepherd, and yet they go astray! Now
listen to this—"*All we* like sheep have gone astray; we
have turned every one to his own way, and the Lord
hath laid on Him the iniquity of us all" (Isaiah liii. 6)—
gone astray like the poor silly sheep, gone our own way

instead of God's way ! (Enlarge and give instances.) *All*
we : not one keeps close to God's way. What are we
to do? say, " Father, I have sinned," and ask Him to
take you back. And what shall we do with the sin, the
terrible sad sin that must be punished—" the Lord has
laid on *Him* "—on *Jesus*—" the sin of us all."

> " I lay my sins on Jesus,
> The spotless Lamb of God ;
> He bears our sin and frees us
> From the accursed load."

Sunday before Palm Sunday.

" SUFFER LITTLE CHILDREN TO COME."

" But Jesus called them unto Him, and said, Suffer little children
to come unto Me, and forbid them not : for of such is the kingdom
of God."—*St. Luke* xviii. 16.

(A thorough explanation of the word " suffer " should
be the first thing in this lesson ; it would even be better
to change the word for the more familiar one of " let " or
" allow," than to leave the impression that it means
" something that hurts," as a little child told me once.)

Do you know that beautiful hymn—

> " I think when I read that sweet story of old,
> When Jesus was here among men
> How He called little children like lambs to His Fold " !

—just like your text, you see.

Why *sweet* story ? one we like so much. We love to

think how kind Jesus was. Suppose the Rector came to your house, talked to Father and Mother, looked very kind and wise and good ; suppose he turned and said with kind voice and loving smile, " Now I want to see the *children*," how pleased you would be! Have heard in other lessons how people came to Jesus and got good from Him—(Lesson 9) healed them, (Lesson 11) taught them wonderful lessons. Some among them thought of their children; your fathers and mothers think of their children, don't they? Father works hard for you, Mother takes care of you, gets you food and clothes, does so many kind things—kindest thing of all to bring you to Jesus ; and perhaps it was not only fathers and mothers —kind Aunt, good Grandmother, or big Sister, all might have helped to bring children to Jesus, just as *Teachers* try to do now. Disciples told them to go away—thought Jesus did not want children to trouble Him. They made a mistake : Jesus looked and called the children, and said—? Jesus so loving, kind, and good. That was long ago; little Jewish children; but Jesus " Same yesterday, to-day, and for ever." Your mothers and fathers brought you to Jesus when you were baptized—ask them to tell you about Him, to help you to come to Him now, by thinking of Him, loving Him, speaking to Him : He will hear you, will say, " Let the little children come."

Palm Sunday.

CHILDREN'S PRAISES.

"All glory, praise, and honour,
To Thee Redeemer, King,
To Whom the lips of children
Made sweet Hosannas ring."

(In picturing scene make *Jesus* central figure, children merely allowed to join, accessories; do not let them have impression of children being most important point.)

Great crowd, people waiting, running to see, looking eagerly—King coming (example, Thanksgiving Day). In Jerusalem, more than 1800 years ago, people waiting, crowd shouting to do honour to some One. Who? A King —King of kings, but came meek and lowly, not in grand carriage, but riding on an ass—still Christ the King. Not earthly king, but heavenly king. Went into Temple, His Father's house—little children there. You remember how Jesus said they might come to Him; what did He say?—"Suffer (let) the little children come unto Me." Now they came to praise Him and sing sweet hosannas. Jesus loved to hear them sing.

We too may sing to Jesus, though He is in Heaven now; specially in God's house. Try to join in singing. Jesus our Redeemer—bought us, freed us from sin; Jesus our King. So we love to join in singing verses like the one we have learned just now.

Good Friday.

CHRIST DIED FOR US.

"While we were yet sinners Christ died for us."—*Romans* v. 8.

Been hearing a great deal about sin and sinners all time of Lent. Who first sinner? what sin? What happened when Adam became a sinner? What happened to sinners in Noah's time? Sin always brings punishment, and punishment for sin is death. Like wages; when people do work they get—? (wages); if good work and good Master then good—? All bad together—bad Master, bad work: what sort?—disobedience, lies, stealing, anger, tempting others to be bad too. All bad work—bad *wages* too: "the wages of sin is *death*." Are we sinners? Yes; *we* say "Father, I have sinned;" *we* say, "All we like sheep have gone astray" (see Lesson 19), so we deserve death for wages. It would be a terrible thought, but there is wonderful news—"*while* we were yet sinners *Christ died for* us."

Tell story of Crucifixion, very quietly and solemnly; do not let children speak, except in answer to careful questions. Took our sins. Whose? every one's—whole world. What sin? all sin, none too great, none too small; took our sin and our punishment. What punishment? And all this while "we were yet sinners." Oh, why do people ever say to children, "Do not do so and

so; if you are naughty God won't love you"? It is
not true, it was while we were yet sinners that God so loved
the world that Christ died for us. But caution must
be used in speaking to the children, as their own mothers
often say the above, meaning that God does not love
sin, which indeed is true.

It is of chief importance that each separate child should
know the text thoroughly in this lesson. Let them have
it to carry with them all through their lives.

Easter Day.

CHRIST RISEN.

"But now is Christ risen from the dead, and become the first-fruits
of them that slept. For as in Adam all die, even so in Christ shall all
be made alive."—I *Cor.* xv. 20, 21.

Is this form that you are sitting on *alive*? can it
move without being touched? can it speak, think, do
anything? Are we sorry it is not alive? No, it never
was alive, never meant to be, more useful as it is: now
think of something that was alive once, not now—little
dead bird, lying helpless, pretty feathers ruffled, sweet
voice silent, bright eyes closed; makes us feel sorry.
More sorry still when we see dear friend dead, little
sister or baby brother perhaps: how sad mother is,
little feet no longer running about, little hands still—
death very sad thing. What was death wages of? Who

sinned first? and we all follow, "As in Adam all die"—very sad; but to-day not a sad day—what day? Very glad joyful day—why? because of the rest of our text, "even so in Christ shall all—"? "be *made alive.*" Did Christ make alive any when on earth? (Describe waking of Jairus's daughter, or widow's son at Nain.) Just so, one day, all shall wake again. What is that day called? Resurrection day—making-alive-again day. How do we know for certain? because Christ is risen. (Repeat first text, "Now is Christ risen, and become the—"? "first-fruits"—Christ first, all to follow some day. Which day? (Tell how Christ was buried, rose again.) What do we call day we keep as day Jesus *died*? How many days ago? Now we keep day He rose again—Easter Day; very joyful, for we can say "*Now* is Christ risen."

First Sunday after Easter.

WHAT THE RISEN LORD IS TO US. I. SAVIOUR.

"Thou shalt call His name Jesus, for He shall save His people from their sins."—*St. Matt.* i. 21.

Is Jesus alive now? Did He ever die? how? What do we call day we keep as day Jesus died? What is done with dead people? They are buried. Was Jesus buried? Is He in grave now? What happened? Rose again; people saw Him, touched, talked to Him—quite sure, same

Jesus, same "yesterday, to-day and for ever;" so *now*,
though gone up to Heaven, same Jesus; and what He
is to His people you are going to be told next few
Sundays.

I. SAVIOUR. What does text say? Jesus means
Saviour in Hebrew tongue—His Name Saviour (see
Lesson 6, same text). Illustrate *saving* by something
familiar to children,—lifeboat going to wrecked ship,
if children know about storms at sea; fireman dashing
into burning house; slave-ship boarded by English,
slaves set free; man or child on railway line; bridge
broken down, brave man crawling over to warn coach
or train. Saved from danger. Do we want saving?
are we in any danger? yes—bad thing inside us, in
our hearts, gets worse unless saved from it. What
did we talk about in Lent?—bad thing Adam began,
worse than storm or fire; that only hurts *body*, sin hurts
soul, separates us from God, not fit to go to heaven.
Cannot save *ourselves*, nor each other; not I you, nor
you me; but what does text say?—the very thing,
"save us from our sins." Came to be the Saviour,
lived on earth to teach people that He was the Saviour;
died to be the Saviour. What did His own Mother
say? (Luke i. 47) "My spirit hath rejoiced in God my
Saviour"—that was Jesus. Where is Jesus now? Is
He the same? Then if he was Saviour on earth He is
Saviour in heaven. Then remember the risen Lord is

I. Saviour—our Saviour to save us from our sins.

Second Sunday after Easter.

WHAT THE RISEN LORD IS TO US. II. GOOD SHEPHERD.

"Jesus said, I am the Good Shepherd."—*St. John* x. 11.

What the first thing our risen Lord Jesus is to us heard last Sunday? Saviour. That not all: when poor shipwrecked people brought in by lifeboat they are safe, but what more do they want? they are wet, cold, hungry; want to be warmed, clothed, fed, led into fresh place. Little child saved from coming train wants to be cared for, led home again. For younger classes, enough to describe sheep as they were in the East, large flocks in open country, wolves and lions ready to eat them, needing shepherd to care for them, watch over, even die to defend them if need be, guide, guard, and feed them.

Older ones may be led step by step through the parable (John x), dwelling on the loving care of the Good Shepherd, *because* the sheep *belong* to Him; not merely hired for money, but each little sheep His very own— longs to keep it safe. On from Jesus *the* good Shepherd to Psalm xxiii, "The Lord is my Shepherd"—no want unsupplied; all our need; each step of the way, right on even to the "valley of the shadow of death"— Shepherd been there before us: when? how? Yes, right down into the valley of death. We need not fear; He will carry us safely through. Story of soldier

on battlefield, left to die alone, found cold and stiff and
dead, but in his hand a little book, prayer-book, and
finger fixed firm in the place of this beautiful psalm,
pointing to words "Though I walk through the valley
of the shadow of death I will fear no evil; Thou art with
me"—happy smile on soldier's face, not alone, Jesus
with him, Jesus Saviour, Jesus Shepherd.

Third Sunday after Easter.

WHAT THE RISEN LORD IS TO US. III. MASTER.

"One is your Master, even Christ."—*St. Matt.* xxiii. 10.

> "Christ is your own Master;
> He is good and true,
> And His little children
> Must be holy too."

Suppose there was some one whom you loved and hon-
oured very much, and you had grieved very much because
you thought they were dead and you would never see them
again; and then suppose you did see them, and you
were able to speak to them—I wonder what you would
say. Will tell you what a woman called Mary said when
she saw Jesus after He was risen. Ask about Jesus's
death, burial, resurrection; tell how Mary came to grave,
very early, stone away—went to tell Peter, came again,
stayed crying—thought the body of Jesus had been
stolen. Had it? Looked again—two Angels; Jesus

Himself—called her by her name. What did she say? "*Master*"—that was what her risen Lord was to her; not merely friend and teacher, but Master. He had bought her with His precious Blood, saved her from her—? she would be His servant, and say to Him with all her heart "*Master.*" *Our* Master too—"One is your Master, even Christ." "No man can serve two masters"— Jesus said that, very true. We have to choose; keep to one, choose the best. Servant going after place, hears of two, tries to find out which is best—best work, wages, home, Master.

We must choose between two—Jesus, Satan.

Satan's work,		sin;	Jesus's work,	being good, doing God's will;	
,,	wages,	death;	,,	reward, crown of life;	
,,	home,	hell;	,,	home, Heaven.	
Master, father of lies.			Master, God of truth and love—Saviour, Shepherd.		

Satan tries to deceive us into thinking we shall find pleasure in serving him; but no, we *never* shall; he will make us slaves, tied and bound with the chain of our sins. Jesus can set us free; He can and will if we ask—free to choose His service, to look up and say to Him in loving earnest tone, "Master!"

Repeat verse of hymn. If Jesus is our Master we must do His work, ask Him to bless and accept it from his little servants.

Fourth Sunday after Easter.

WHAT THE RISEN LORD IS TO US.   IV. LEADER.

"Christ also suffered for us, leaving us an example, that ye should follow His steps."—I St. Peter ii. 21.

What do men want when a party of them join together to do anything? a leader. Bring this out by questioning; show how useless expeditions would be if each man followed his own way, how necessary to success that all should work under a leader. Israelites made great expeditions, great journey from—? Egypt; to—? Many difficulties. Who was their Leader? Psalm lxxvii. 20. God Himself led them (cloud and pillar of fire) by the hand of—? Moses and Aaron; later on, Joshua; by and by, David (Isaiah lv. 34).

With very little children the game "Follow my leader" might serve as illustration instead of Bible examples—one chosen leader to run about field or common or playground, all the others follow, do as he does; he is leader because he leads them; they follow, go after in his steps, try to be like him, do as he does: this only game, but in same way for useful things *leader* wanted.

Israelites wanted leader : (1) Because they were journeying—where from? where to? (2) Because they had to fight.

We have same reasons—journey of life; life never

stands still, always going on ; in journey every step makes
more way behind, less in front : so with our lives, each
day one more behind, one less in front. Do we want
the end of the journey to be Heaven ? then we must have
the Leader, Who can take us there, Jesus our Leader.
He is the way. Remember your hymn—

> " Only Thou our Leader be,
> And we still will follow Thee."

They had to fight; so have we—made soldiers of
Christ at our Baptism (see Baptismal Service). All
know hymn " Onward, Christian soldiers " ; look at lines
6 and 7—

> "Christ the royal Master,
> Leads against the foe."

" Master " last lesson, Leader now—our Leader. Oh,
let us faithfully follow Him, step by step, day by day—
Clergyman put in his study, " What would Jesus do ? "
Let us put it in our hearts, and try to follow it.

Fifth Sunday after Easter.

CHRISTIAN CHILDREN.   GOD THE FATHER.

Hymn, verse 1—

> "We are little Christian children;
> We can run, and talk, and play.
> The great God of earth and heaven
> Made, and keeps us every day."

When you go away out of school, suppose you go down
such and such streets (describe way to church), what do

you see? large building (describe), what is it? And if you go into next parish you find another church, next town more, far away down in country still churches; no city, town or village without church all through England, why? because England a Christian country. The churches are for Christian people (part of the great body of the Church of Christ); and you are *Christian* children, made Christian when baptized. What does Catechism say? made "member of Christ." Grand thing to be—hold fast to it, thank God for it. St. Peter tells us that even to *suffer* as a *Christian* is a happy thing (I Peter iv. 14, 16). Ethiopian so glad to be baptized, made a—? went on his way rejoicing (Acts viii. 39). Jailer too rejoiced (Acts xvi. 34), he and all his house; so we think perhaps little children there. Happy children to be made Christians! And we *know* of some children brought to Christ Himself: what did He say? Happy children to be let come to Jesus! May you come? Now if you are Christians you have got to believe something; jailer had; was asked before he was allowed to be baptized; you were too little to be asked, but people who brought you asked, and promised you should believe all the Articles of the Christian faith (see Catechism). What first Article?

"I believe in God the Father Almighty"—God the Father of Jesus, Father of all; almighty, able to do everything, made me and all the world; great good God, ever-living, all-seeing, made whole world; made

us and attends to us, "keeps us every day." How—?
breath, air, light, food, warmth. Holy God; hates sin,
but loves us; will forgive through Jesus, Who teaches us
what little children can learn about God.

God is love; and God our Father.

Sunday after Ascension.

GOD THE SON.

Hymn, verse 2—

> "We are little Christian children;
> Christ, the Son of God most high,
> With His precious blood redeemed us,
> Dying that we might not die."

Recapitulate last lesson about *Christian*; work out that
it is a happy thing to be a Christian, and then lead up
again to Articles of Belief.

First article?—stand and repeat reverently.

Second?—"I believe in God the Son, Who hath redeemed me and all mankind."

Those who can repeat (or read) Apostles' Creed may
do so: then recapitulate, following step by step, Christmas lesson, "born of the Virgin Mary"; Good Friday,
"crucified, dead, and buried"; Easter, "third day rose
from the dead." "Ascended into heaven," lesson for
Ascension Day, last Thursday; keep it holy, happy day;
go to Church. Describe first Ascension Day (Acts

E 2

i. 9, Luke xxiv. 50); disciples watched Him go up, where
to? In Heaven still? Look at Creed again, " sitteth on
right hand of God." We cannot see Him. St. Stephen,
who afterwards died for Jesus, saw Him in Heaven once.
Will Jesus ever come back? Creed again, " From thence
He shall come." Are we getting ready for Him (Advent
lesson). Disciples went back to Jerusalem with joy—
why? they had lost their Master, before, on Good
Friday; they were very very sorry, now glad! *Then*
thought He was *dead*, now knew He was alive, and would
never die, would come again—busy and happy getting
ready. If we are real true Christians, believing in
the Christ Who redeemed us, by being born a child,
living, dying for us, rising again, waiting at God's right
hand in Heaven—then we can be happy like disciples,
can go and do the work Jesus gives us to do with glad
and joyful hearts till He comes again.

Disciples had another reason too for being joyful—
Jesus gave them a promise; He knew they would have
many troubles and difficulties, could not be good Christians and follow Jesus by themselves; so promised that
God the Holy Ghost should come into their hearts, to
teach them, and speak to them of Jesus (John xiv. 27;
xv. 26).

Whitsun-Day.

GOD THE HOLY GHOST.

Hymn, verse 3—

"We are little Christian children;
God the Holy Ghost is here,
Dwelling in our hearts to make us
Kind and holy, good and dear."

Children should repeat this very softly, and reverently.
Teacher can best ensure this by example; in every case
where reference is made to God the Holy Ghost, or to
the sufferings and death of our Lord, to mark reverence
and feeling in Teacher's tone is far more efficacious than
the often repeated exhortation "You ought to feel solemn."
True reverence will be readily caught by the children,
and it is especially necessary to set right tone when teaching a subject so mysterious and full of awe as this.

Recapitulate Ascension lesson—Jesus gone, His people
left on earth, but not to be left to bear their trials and
difficulties alone; What was the promise? Who would
come? Jesus gave Him such a sweet name, "Comforter"—to be help, comfort, strength to them. (Para-
clete = more than our word Comforter.)

Promise came true: to-day we think specially about
it; disciples did not have long to wait; Jesus first told
them before He was taken by the wicked men and put to
death, before—? Good Friday. Then what happened?
rose again. When? Then forty days Jesus Himself

with His disciples, went up to heaven on Ascension Day, Thursday last week, now Whitsun-Day; how many days on? ten. 40 and 10 make—? So it came to be feast of Pentecost, *pente* = 50—and then promise came true.

Could disciples see God the Holy Ghost when He came? No, Ghost means spirit, "God is a Spirit"— cannot see a spirit, like wind; *feel* wind, see what it *does*, not see it. Disciples could see Jesus when He was on earth because He was made man; Holy Ghost not made man, came a Spirit into their *hearts*; they could *feel* good thoughts come, feel help, comfort, strength. So with us—part of our Creed "I believe in God the Holy Ghost, Who sanctifieth me and all the elect people of God." Great gift at our baptism.

Repeat verse of hymn. Do not let us grieve Him (Eph. iv. 36). Listen to His voice in our hearts, and obey.

Trinity Sunday.

GLORIA.

Hymn, verse 4—

> "We are little Christian children,
> Saved by Him Who loves us most;
> We believe in God Almighty,
> Father, Son, and Holy Ghost."

Whole hymn may be sung or repeated.

When lesson ready to begin, children stand, repeat Gloria Patri all together, reverently, distinctly.

This Trinity Sunday lesson must be short: not possible to keep children long in state of reverent awe desirable for the subject of God's glory.

Try to lead classes as far as you can, then break off before they individually stray, and repeat hymn or talk of some of the beautiful things that help to show God's glory on earth—sun, stars, flowers, trees, gold, bright colours; or the glories of Heaven.

Leading thoughts for lesson:—Sun in its power very bright, glorious; but nothing compared to glory of God. Told about God in—? Bible. We may not see His glory—too sinful; but we believe, and worship, and say, "Glory be to the Father, Son, and Holy Ghost." Been learning about all they have done for us.

> Father—   made me, etc.,
> Son—   redeemed;
> Holy Ghost—sanctifieth.

Now to-day give glory to Three in one—Trinity: "As it was, is now, ever shall be." Abraham, Moses, David, all knew God's glory as it *was*: same now; though we cannot see, sun, trees, flowers, all help to give us thoughts of God's glory, and God's Book. "Ever shall be"—up in Heaven, where we shall see so bright, so wonderful. We shall praise God better then. Who praise God in heaven? Angels—never tired. Let us pray God that we may join their praise, with white garments washed in Jesus' blood, and hearts made holy by His Spirit.

First Sunday after Trinity.

WIDOW'S SON AT NAIN.

"Jesus said, I am the Resurrection and the Life; he that believeth in Me, though he were dead yet shall he live."—*St. John* xi. 25.

Learn and say these words very carefully, quietly, reverently; part of Burial Service of Church of England; beautiful words said by Jesus about Himself and those who believe in Him; like them to be said about us when we die. Very sad to see funeral! sometimes it is a father, or mother, or sister, or baby: people in black dresses, cry, very sorry; but clergyman comes, says these words full of comfort. (Repeat.) Remind of Easter lesson, "In Adam all die, but in Christ all made alive." Wonderful thing for Jesus to do. Are we sure He can? Two reasons—first, He is risen Himself, "become the first-fruits." Easter lesson again—Christ first, His people follow. Secondly, we know He can, because he made people alive when on earth. Said how sad to see funeral; hope you never stand and stare to see it go if there is one near your house, never laugh or point or crowd round, as some thoughtless people do. In some countries men and boys always lift their hats, and all should show respect; speak gently when speaking of any one dead, or telling of death.

Jesus met funeral one day (Luke vii. 11). He was going into little town, Nain, met company coming out—

so sad, poor widow mother, her only son dead; no
husband, no son; people very sorry for her, came with
her to bury him; Jesus sorry for her too. Friends
could not help, Jesus could—came and touched the
coffin, spoke to the dead man. Could he hear? yes,
even the dead hear the voice of Jesus. What did Jesus
say? "*I* say unto thee, Arise." He could obey when
Jesus spoke. Jesus is God; the winds and waves obeyed
Him; the dead obey Him. The young man arose, and
Jesus gave him to his mother—what a precious gift?

What was this wonderful thing (miracle) Jesus did?

Shall we hear Jesus's voice even when we are dead?

Mother's or nurse's voice calls us and wakes us out of
sleep—Jesus will wake us Resurrection morning.

Second Sunday after Trinity.

PETER'S WIFE'S MOTHER.

"He giveth power to the faint; and to them that have no might
He increaseth strength."—*Isaiah* xl. 29.

Who is this? God. True of Jesus too, for Jesus
is God the Son. Do you remember verse (Luke iv. 40,
Lesson 9) when sick people were brought to Jesus one
evening, and He healed them all? Was not that
giving power to faint and strength to them that had
no power. Going to hear about a sick person to-day.

St. Mark i. 30; Holy Land, place called Capernaum, sea-
side, fisherman's house; belonged to two fishermen,
Simon Peter and Andrew, both disciples of Jesus.
Some one else lived there too, or at least was there the
day I am speaking of—Simon's wife's mother; old
lady; if there were any children would be their grandmother. Have you got a dear Granny?—love her very
much, be kind to her, do things for her if you can. It
was Sunday, and great thing going to happen in that little
house—Jesus coming to it! had been to synagogue
(church), came home with Simon and Andrew, brought
two more, James and John; only one thing to spoil the
great pleasure and honour Simon Peter felt—mother-in-
law very ill in bed, got bad fever. Ever seen any one
with a fever? very bad thing—hot, toss about, very thirsty,
very ill; and even if fever goes, very weak long time
after—like people in text, faint, and no power. Did
not tell Jesus at first; perhaps thought He would not
come if He knew, would be afraid of giving extra trouble.
But Jesus did know; knows all things, knew He could
help, not trouble. Did tell Him after a while. What
did He do?—go away? no—that would be kindest
thing to do if any one could not help: but Jesus could—
went to bedside, took her hand, lifted her up, told fever
to go away. Did it obey? yes, He is God. Then did
first what our text says. (Repeat.) Instead of the old
lady being faint and weak she was now strong, able to
get up and work: Jesus gave her power and strength—

instead of being in bed, waited on by others, not able to feed herself, now she walks about, and at once begins to get dinner ready for Jesus and disciples, to do all she could for them. I expect she wanted to do all she could to show love and thanks to Jesus for making her well. Has Jesus done anything for us? Do we want to do all we can for Him?

Third Sunday after Trinity.

JAIRUS'S DAUGHTER.

"Be not afraid, only believe."—*St. Mark* v. 36.

Heard about Jesus raising up the widow's dead son. What was the name of the place? Was the widow very sorry when her son died? Friends sorry too, went with her to funeral. Who met them? Jesus sorry for her too. What did He do?

Another wonderful story of something Jesus did—not anybody's son this time, but a little girl; father's name Jairus, great man, ruler of the synagogue, but very sad just now, little daughter very ill, twelve years old, going to die, doctor could not do any more for her, no hope. What a sad house when that is said! father loved her very much, very likely had done everything he could for her—best doctors, medicine, all in vain. Could not bear to see her die; thought of one thing more he could do—go and find Jesus; had heard how He made sick people well. Can

you tell me of anybody? Peter's wife's mother, last
Sunday lesson; and then all the people in the evening.
Can you repeat verse? Luke iv. 40.

Perhaps thought of that gave Jairus hope that if he
could only find Jesus, beg Him to come, his little girl
might be made well. Read verses 22, 23, part of 24.
And Jesus went with him. How glad the poor sad
father must have been when he found that Jesus was
really coming with him! Where did he find Jesus? By
the seaside. What did he do? Where is Jesus now?
Can we find Him? Can He hear us speak? Then when
we want Jesus to do anything for us, we must ask,
as Jairus did; kneel down, beg very earnestly.

Not end of story yet: something happened—great
crowd in the road, great many people pressing round
Jesus, delay, waiting to speak kindly to sick woman.
Very hard for Jairus to be patient, wanted so to get
home; why? Jesus knew it was hard, but knew that it
was a good thing—would give him more patience, more
faith; could do more for him by waiting. So sometimes
with us: our prayers not answered at once—wait and
trust Jesus. Now came very sad thing—messengers from
Jairus' house, bad news, the little girl was dead, too late!
No use to trouble Jesus to come now, they said. Jesus
was speaking to poor sick woman, but He heard; knew
how bad the poor father would feel, what great trial for
his faith; made haste to say the good words of our text
" Be not afraid, only believe." (Repeat.) Knew He

could help still, wanted poor sorrowing man to know it too, to *trust* Him. Can you wait till next Sunday to hear what Jesus did? Remember and trust Jesus. Be not afraid, only believe.

Fourth Sunday after Trinity.

JAIRUS'S DAUGHTER (*continued*).

"Damsel, I say unto thee, Arise."—*St. Mark* v. 41.

Recapitulate last lesson—father's name, who was he? why sad? Who did he go to look for? Where did he find Jesus? Why did they not get home quickly? What bad news came? What did messengers think?—no use to trouble the Master (who was the Master?) any further. Did Jesus think so? What did He say to Jairus? Last Sunday's text, to-day's, tells us what Jesus said to the little girl.

Like to hear rest of story? Left them all in the way, Jesus, the ruler, messengers, crowd went on; but going now to house where some one was dead. Who? Crowd not wanted there; only Jesus and Peter, James and John, allowed to see what wonderful thing (miracle) Jesus was going to do. Reached the house, came in. Anybody there? Little girl's mother there: no other daughter to comfort her, very sad! no hope! no comfort! She did not think Jesus could do anything now—too

late! most likely it was she who sent the messengers to
stop His coming, did not wish to waste His time and give
Him trouble for nothing. Was it for nothing? We
shall see : mother so sure that little girl was dead and
would have to be buried, that she had let the people
come to see about it, to weep and wail after the Jewish
custom. Jesus did not like all this noise and confusion,
said (verse 39) " why? the damsel is not dead but
sleepeth." Damsel = little girl, maiden. Christian people
often said to sleep when they die (1 Thess. iv. 14), " sleep
in Jesus;" 1 Cor. xv. 20, "first-fruits of them that
slept;" why? because they will wake again. Little
sleep every night, wake in the morning; long sleep of
death, wake Resurrection morning, when Jesus calls us.

Another reason for saying this of Jairus's daughter,
she was going to wake very soon, because Jesus called
her soon, took her by the hand, said words of our text
(repeat)—" Little maid, get up, Jesus tells you to arise."
Did she hear? yes, though her body was dead, her
spirit flown away, she heard Jesus, obeyed Him at once.
Wonderful thing that the dead could hear Jesus' voice
and obey Him! What are the wonderful things Jesus
did, called? Miracles. And Jesus could do miracles
because He is—? God.

Fifth Sunday after Trinity.

BLIND BARTIMÆUS.

"And he cried, saying, Jesus, Thou son of David, have mercy on me."—*St. Luke* xviii. 38.

What are the wonderful things Jesus did, called? miracles. Hear about miracle to-day—Jesus journeying on towards Jerusalem, walking with His disciples, and great company of people wishing to hear and see: why hear? because "He spake as never man spake;" see? because of His miracles; and some because their hearts were touched with love for Him. Got near to Jericho; somebody sitting down by roadside, blind man; blind not taught to work in those days—if poor, could only beg; very sad and helpless! Could he see people passing by? no, but blind can hear, very quick ears generally; heard noise and stir of crowd, asked somebody what it was (verse 37); and they told him "Jesus of Nazareth passeth by"—Nazareth (Matt. ii. 23) place where Jesus lived, so called "of Nazareth," as we might say "of Bloomsbury," "of York." Has the blind man ever heard of Jesus, do you think? yes, evidently—for at once he began to do something, what our text tells us (repeat); like Jairus, you see, wanted something, began to ask, to pray; called out loud. People near him, those coming on in front, said Hush, did not like to hear him, did not want Jesus to hear; blind man *did* want Jesus to hear, so cried out more than ever, same words, "have mercy on me." Picture scene that followed (verses 40–43)—Jesus standing in midst of waiting multitude, blind man sent for, brought face to face with Jesus; faith is growing, hope dawns; wonderful question, "what do you want Me to do?"—Give me sight. The answer. Do you remember when God said, "Let there be light, and there was light"? like that now for blind man— gives praise to God, he and all the people.

We too, like blind man, want mercy, pity and help from Jesus. We too can be heard by Jesus, though we cannot see Him. We too must have faith and pray.

Then when Jesus says, "What do you want me to do for you?" we can tell Him what we want.

(Lesson may be enlarged by reference to spiritual sight, Ps. cxix. 18, "Open Thou mine eyes, that I may see won- drous things in Thy Law," and "From all blindness of heart, Good Lord, deliver us."

Sixth Sunday after Trinity.

SERMON ON THE MOUNT. I.

"And seeing the multitudes, Jesus went up into a mountain: and when He was set, His disciples came unto Him."—*St. Matt.* v. 1.

Read chap. iv. 23, 24, 25.

Going to hear about very wonderful sermon preached

by Jesus more than 1800 years ago. None of us there to hear it; how do we know about it? from the—? Bible.

Now if we were not there, who was there? Was anybody there? The disciples. Who besides, few or many? And "seeing the *multitudes*"—is that few or many? many, great many (verse 25). Where did Jesus go to preach from? Up into a mountain. Clergyman in church goes into a pulpit; to see the people, let them hear well; Jesus had high hill for pulpit; and now where did people come from? (verse 25) long, long way some of them, many miles—why? To hear Jesus. What made them want so much to hear Jesus? (verses 22, 24). Because of what He had done. Sometimes we hear people say, " Oh that's the clergyman that's been so kind to me, who came to see my mother when she was ill; I want to go and hear him preach." They want to hear what he is going to say because of what he has *done*. Jesus had done far more than any man could do (Jesus God too), and of all the people that followed Him such a long way most of them could tell some good thing He had done—" made my wife's. mother well of a fever," St. Peter could say; and many other people had had themselves or their friends made well; that made them want to hear Jesus, for three reasons :—

1. He was great—could do things nobody else could ; so could tell them things nobody else could about God.

2. He was good, or He would not have cared to help

F

and heal them—so He would tell them good things about
God loving them.

3. They were grateful to Him, and wished to show
Him that they were so.

We shall hear next Sunday what Jesus said to them.
Now I want you to think. Have not we got the same
reasons for hearing Jesus? Those people could tell of
things Jesus had done for them; so can we. And one
great thing they did not know. He died for us; we
know He is great and good; shall we not be grateful,
and listen to Him when He speaks by His ministers,
His Word and His Spirit?

Seventh Sunday after Trinity.

SERMON ON THE MOUNT. II.

"And He opened His mouth, and taught them, saying, Blessed
are the poor in spirit: for their's is the kingdom of heaven."—(Com-
pare Psalm xxxiv. 18.)

Recapitulate last Sunday's lesson. Where was Jesus?
why? Going to preach. Who to? The people who had
followed Him so far. Who there besides multitudes?
Disciples.

Now what did Jesus say? You are going to learn
only a part of the sermon, a very beautiful part. Jesus
did not begin by scolding the people, or by talking about

their sins, though they were sinful, and He had to tell
them of their sins sometimes; but now He begins by
telling them who are the happy people. "Blessed"
here means "made happy by God" (Prov. x. 22). We
might say "*happy* are the poor in spirit, for theirs is the
kingdom of heaven."

What sort of people are the poor in spirit? and
why are they blessed? We call people poor when
they have not got things that others have—poor in
purse when they have not got money, poor in body
when they have not got health, poor in spirit when
they have not got pride in their hearts.

Jesus was sorry when He saw people poor in body,
He healed them; He liked to go among people who
were poor in money. He had come on purpose to bring
them good news, and He knew that very often being
poor in money was good for them, made them more
likely to be what our verse says "poor in spirit" too.

God blesses the poor in spirit (Ps. li. 17; Prov. xvi.
19); He "resisteth the proud" (James iv. 6); and Jesus
wanted all these people who heard Him to please God
and have His blessing by being poor in spirit. Illustrate
by parable of Publican and Pharisee (Luke xviii. 10):
Pharisee proud and haughty, thinking of all his own
good deeds, not of his sins; Publican had not got the pride
and the haughtiness; he was—? "poor in spirit."

Does Jesus speak these words to us? Yes, by the
Bible; we said we were to listen when He spoke; now

we must remember, if we want to get the blessings
we too must be poor in spirit—not proud, not thinking we are good when we know that every day we do
bad things, and all our good comes from God. God so
high above us, so great, so good, so holy, we can only
just humbly say, like the publican, "God be merciful
to me a sinner."

Eighth Sunday after Trinity.

SERMON ON THE MOUNT. III.

"Blessed are they that mourn, for they shall be comforted."—
St. Matt. v. 4.

How strange this verse seems at first! different from
what people used to hear and think. Happy those that
mourn. What does "mourn" mean? sad and sorrowful.
Do people like to be sad? no, but often have things to make
them so—disappointed of what they hoped to have, or to be
or to do; ill or poor; or those they love die or go away:
and worst of all, sin makes people sad. Every one in the
great multitude knew what it was to mourn, and every one
in the great world now does—even our good Queen not
free from sorrow; but now told happy to be a mourner!
why? "for they shall be—"? "comforted." How sweet
that is! little children comforted by their mother, so will
God do for us (Isa. lxvi. 13); Jesus came on purpose (Isa.
xi. 2) "to comfort all that mourn." He knew He could

do it, that was why He said, "Happy are those that mourn."

He came to give *life* to the *dying* (John x. 10), to *find* those who were *lost* (Luke xv. 4, xix. 10). He came that the *weary* might have *rest* (Matt. xi. 28), *dead* have *hope* (John xi. 25), *sin* be *destroyed* (Rom. vi. 6).

So that in *every* trouble Jesus can give comfort. People had new thought given them—"better to mourn with Jesus to comfort you, than to try to be happy away from H m." True now. Sweet message you can give when you see people unhappy—whisper to them that Jesus said, "Blessed are they that mourn, for they shall be comforted." Remind them to ask Him to do it for them. And remember it for ourselves when we are in trouble—look to Jesus to comfort us.

If desired, the lesson may be expanded, by taking the point—better to mourn now and be comforted by and by, than worldly pleasure now and sorrow in future, illustrated by parable of Dives and Lazarus (Luke xvi. 19).

Ninth Sunday after Trinity.

SERMON ON THE MOUNT. IV.

"Blessed are the meek, for they shall inherit the earth."—*St. Matt.* v. 5.

Recapitulate last lesson. Whom did Jesus call blessed? Why? Who will give comfort? Who next called blessed?

—the *meek*. What sort of people?—those who submit quietly to God's will, listen to His Word (James i. 21), are not easily provoked or put out of temper, not loud and noisy, always talking of their rights; they leave God to fight for them, and He does (Ps. lxxvi. 9). St. Peter calls it the "ornament of a meek and quiet spirit," says that in the sight of God it is of great price, very precious and beautiful. It was not a *new* lesson the people heard this time; had heard the very words before (Ps. xxxvii. 11). Many blessings promised to meek in Old Testament—that they should eat and be satisfied (Ps. xxii. 32); that God Himself would guide and teach them (Ps. xxv. 9); would lift them up (Ps. cxlvii. 6); would beautify them with salvation (Ps. cxlix. 4); that Jesus would preach good news to them (Isa. lxi. 1) (just what He was doing in the sermon); make them more joyful through Him (Isa. xxix. 19). And once, in a terrible time of God's anger, the prophet (Zeph. ii. 3) called upon the people to be meek, for if the meek should seek the Lord may be He would have mercy upon them.

Now would not you think that with all these beautiful promises everybody would be meek? Ah, but we are all so sinful, only one man in the Bible said to be very meek—Moses (Num. xii. 13). Look at yourselves, are you meek? do not you find you are impatient when you have to do what you don't like, angry when others provoke you, ready to say at once, "That is mine, I ought to have that, to keep this, to go there"? Oh, how often

we hear the angry word or see the clouded face even
here in this room, and on these happy Sabbath days—
"that place is mine," "I don't want to sit next her."
How can we learn to be meek? Jesus Himself will teach
us (Matt. xi. 29): "Learn of Me, for I am meek and
lowly of heart." Jesus our Teacher, our Example.

Tenth Sunday after Trinity.

SERMON ON THE MOUNT. V.

*"Blessed are they which do hunger and thirst after righteousness,
for they shall be filled."—St. Matt. v. 6.*

Repeat from beginning of chapter; then verses 1, 2,
and new verse 6. Who taught this? Who to? Why
called Sermon on the Mount.

Did you have any breakfast this morning? enjoy it?
why? because hungry and thirsty. Do you ever run in
from school and say, "Please mother, piece of bread—
drink of water"? why?—hungry and thirsty. All know
what that means. Jesus even knew—thirsty by well at
Sychar, hungry when He came to Jerusalem. People in
the Bible knew. Can you tell me of a boy? (Ishmael,
Gen. xxi)—no water, nearly died, so thirsty; and of men
who came a long way to buy corn for little ones at home
—so hungry (Gen. xlii). Israelites hungry (Exod. xvi),
and thirsty (Exod. xvii), in the wilderness—David (1
Chron. ii. 17)—two beautiful verses Ps. cxxxvii. 6, 7, that

say what all these did, "Hungry and thirsty, their soul
fainted in them; and they cried unto the Lord in their
trouble, and He delivered them out of their distress."
Say it after me. Would you do that (cry to God) if in
distress and had no food?

All this time talking of bodily hunger: must have
food, to keep our bodies alive and strong; but what
has God given us besides bodies?—a living soul (Gen.
ii). Souls want food too—good healthy bodies hungry
for bread and thirsty for water, good food and healthy
drink; but sometimes bodies hungry for foolish things
and bad drink. Souls may be hungry; i. e. want very
much, long after, foolish things too—worldly pleasures,
"pomps and vanity," or even sin. Oh sad to be hungry
after sin! Let us turn to right kind of hunger for our
souls; what Jesus says people are blessed for having
(repeat text) hunger after?—righteousness, doing right,
being right; holiness of life. They shall be—? What
is that we sing in Church, "He hath filled the hungry
with good things." Jesus knew it would be so, was Himself Shepherd to feed His people. When you are older
you will learn and understand more of the "strengthen-
ing and refreshing of our souls" by Christ; how we do
truly "feed on Him in our hearts by faith," how He is
the "Bread of life," and the "Living Water."

How can we get right kind of hunger? What makes
us hungry in our bodies?

Firstly, given us by God, sign of health; sick people

not hungry : next get hungry by hard work—so with souls, first God's gift, and we must ask for it.

Secondly, we must work for it. To be hungry for proper good things, must keep ourselves from bad things (child eaten too many sweets, not hungry for dinner). Pray God deliver us from evil.

Thirdly, can get more hungry after righteousness by thinking how good and beautiful it is ; by being with good people and seeing how nice it is to be like them ; by reading good books and most of all Bible. Just as you get more hungry by looking in baker's shop, or seeing table spread with nice things. May not go into shop and eat without paying ; but Christ calls our hungry souls to be fed without money (Isa. lv. 1). Think more and more what Jesus was ; how good, how holy. Do you not long to be like Him ? The more you think and long, the more hungry and thirsty for righteousness you will be ; and the more happy, for you shall be filled.

Eleventh Sunday after Trinity.

SERMON ON THE MOUNT. .VI.

"Blessed are the merciful, for they shall obtain mercy."—*St. Matt.* v. 7.

Repeat verses 1 and 2, and new one 7—

Merciful—those who show pity and compassion. Hear about unmerciful man, no pity (Matt. xviii. 23). King, servants, reckoning ; one owing very great sum of money,

much more than he could ever pay; if everything he had
was sold, not enough. All as nothing to the great debt
—ruined, he and wife and children; very miserable.
What could he do? fell down and prayed King his master
to have mercy on him. Master was merciful, forgave him
all that great debt, let him go; was not that good? do
not you think man would be very grateful? You shall
hear—went out, met another man, own fellow-servant
who owed him a little debt, seized him roughly, said,
"pay me that thou owest;" fellow-servant asked him
to have patience and he would pay: no, no patience, no
pity; sent poor man to prison. Was that being merciful?
Other servants went and told the King. He was angry,
sent for unmerciful man, said (read verses 32, 33);
punished him as he deserved. This story told by Jesus
to teach same lesson as verse 7 in Sermon on Mount:
merciful king like God; we owe Him everything, cannot
pay, all spoilt by sin. He forgives us; we must forgive others. What do we say in the Lord's Prayer?
" Forgive us as we—? "

If we are merciful we shall show mercy to—

(1) Any who have hurt us in any way, taken our
toys or books away, told unkind tales of us. God forgives
us, we must forgive others.

(2) Any who may be smaller or weaker than ourselves; boys to girls, big children to little; young strong
people to old weak ones; well to sick, blind, lame, deaf;
people with sense to poor idiots; rich to poor.

(3) Animals—horses. Some of you boys have to do with horses in the Mews now, may be coachmen or driver some day; show mercy to your horses. Cats dogs, birds, even flies on the window, and mice in the cupboard; if must be killed, do it kindly and quickly: never tease or be cruel to animals—God's creatures; pets, don't forget to feed and clean them, or water plants. Be merciful, and you will obtain mercy.

Twelfth Sunday after Trinity.

SERMON ON THE MOUNT. VII.

"Blessed are the pure in heart, for they shall see God."—*St. Matt.* v. 8.

Repeat verses 3, 4, 5, 6. Who said these words? when? how? What does "blessed" mean? What was verse last Sunday? What does merciful mean? Was the King merciful? How did He show it? Was the servant merciful? Do we need mercy? Then we must show it too? (I, II, III).

Verse to-day not to do but to *be* (repeat verse 8).

"Pure" means clean and true; nothing dirty or spoilt —like clear blue sky and little white fleecy clouds on summer day, before dirty black smoke spoils it; like driven snow on mountain top, where no foot can tread; like bright gold freed from dross in the furnace: these all pure to our eyes and thoughts, but even these are dim and clouded to the pureness of heaven.

Verse 4, new thought for people (repeat 4). Was verse 5 new? (see Lesson 4). Verse about pure in heart not new either. David knew what sort of people would see God (Ps. xv. 2, xxiv. 4). Isaiah gave beautiful promise that those who kept evil from going into their hearts, who stopped their ears and shut their eyes from evil (Isa. xxxiii) should "see the King in His beauty." We long for this great joy; and if we become pure in heart, indeed we may see God. If we had the chance to go and see Queen Victoria, should we like to be turned back at the Palace gate dirty and unfit to go in? We should wash and get ready, spare no pains; and how much more we need to prepare our souls to meet God.

How can we become pure in heart?

(1) "The blood of Jesus Christ cleanseth from all sin." Say David's prayer, Ps. li, "Wash me, and I shall be whiter than snow."

(2) Ask God to help you; try very hard to keep from bad things, to be always busy with good ones. What prayer, two lessons ago, to be kept from bad food for our souls? "Deliver us from evil." Bad companions out in the street, oh keep away from them, go indoors, give up your play rather than stay with bad boys or girls. (Teachers press this home: how urgently needed only those know who have watched children becoming evil-minded from contact with bad companions at play in

London streets.) If bad words are said, stop your ears, as Isaiah said; shut your eyes to bad dirty deeds, turn away, each one you give way to will make it harder for you to be pure in heart. What was promised in our baptismal vow?—renounce the devil and all his works, all bad evil things. Turn to Collect (18th after Trinity)—"pure hearts and minds to follow Thee:" that is the way—follow Jesus, He our Leader, keep looking to Him, and ask for Holy Spirit in our hearts to help us to be holy and pure like Him.

Thirteenth Sunday after Trinity.

SERMON ON THE MOUNT. VIII.

"Blessed are the peacemakers, for they shall be called the children of God."—*St. Matt.* v. 9.

Repeat verses 7, 8, 9.

What promise to merciful? to pure in heart? to peacemakers?—Children of God; God our Father; Heaven our home. God is *Love*, so a *loving* Father to His children; Almighty, so a strong Father; Everlasting, so will never die; Unchangeable, so never leave or forsake His children. All good things are His, so He can give good gifts to them. Happy those who are children of God. Let us see why He loves to have peacemakers as his children.

God is called the God of Peace (Rom. xv. 33), for He

loves to give peace in the hearts of His people through Jesus. Jesus called the "Prince of Peace" (Isa. ix. 6), for He came to make peace between God and man. What did the Angels sing on Christmas morning? "Peace and good will to men." This peace is in our hearts. Remember storm on Galilee lake—great waves, boat tossed about; Jesus spoke, "Peace, be still;" all quiet, peace came, a great calm; so when storm of passion rises in our hearts, angry thoughts, bad feelings, let Jesus speak; the angry passion will die away, peace will come, and all will be calm and still.

Another kind of peace where you can be peacemaker—first, in your homes, and amongst your companions, when other people are angry or vexed. Is there any quarrelling in your games or your work? then you be a peacemaker. How?

 (1) By asking God in your heart to make everybody at peace.
 (2) By soft gentle words (Prov. xv. 1).
 (3) By kind loving looks—sunshiny looks are catching.

Secondly, in your country. This more particularly when you are older. What is the name of our country? We love our country and our Queen, we want to see England prosperous and happy, peaceful;—must do our part by not rebelling against the laws, not making game of the police, but respecting authority (see Catechism)—never join in riot and disorder, but help to make and keep peace. Thought to help us—the promise itself; if we are children of God, other people our brothers; they too say, "Our Father." Let us love as brethren.

***

Fourteenth Sunday after Trinity.

SERMON ON THE MOUNT.   IX.

"Blessed are they which are persecuted for righteousness' sake: for theirs is the kingdom of heaven."—*St. Matt.* v. 10.

Repeat verses 1–10. Then verses 1, 2, and new verse. "Persecute" means to follow after to do harm; "persecuted for righteousness' sake," followed after to be hurt because they were righteous. And these people called by Jesus "happy." Yes, and St. Peter called them happy too after he had tried what being persecuted meant (1 Pet. iii. 14). How strange! the people had not heard this before, and it was a long long time before they could believe it; knew Jesus would have a kingdom, thought it was here on this earth, and that they would soon see Him crowned King. Shall we ever see Jesus crowned King? yes, when He comes again. Do we not pray, "Thy kingdom come"? Who will be in His Kingdom?—those who have been persecuted for righteousness' sake. St. Peter, St. Paul, St James, killed for teaching about Jesus; and many other holy men, and women and children too. You will perhaps read about them when

you are older. Jesus knew it would be so; wished to
prepare them. He knew it would be bad, but He would
be with them to comfort and help them to bear it—all
seem nothing when they got to the glorious Kingdom.
Did they deserve to be persecuted? No, done by wicked
men; but they were patient and brave, looked on, up to
Christ and the Kingdom. Better to do right and suffer
for it, than to do wrong and not suffer.

Two thoughts for us:—

(1) We very seldom have great persecution for
righteousness' sake—let us be thankful. Christians used
to live in fear, punished if they came to Church, even put
to death; now everybody glad to see you go to Sunday
School and try to be good.

(2) When you get older you may know a little bit
what persecution means—from bad companions in work-
shop or in service; you may be teased and laughed at by
bad thoughtless people, just for trying to do right, for
saying your prayers, reading your Bible, or coming to
school or church; or perhaps for refusing to cheat or
tell lies. Remember then what you are learning now;
bear it bravely and patiently; think of Daniel, how brave
in saying his prayers; think of Jesus Himself, how
patient, though He was persecuted more than any man.
He was God: could have put an end to it at once—why
did He not? Because He chose to bear it for us, to die
for us.

Fifteenth Sunday after Trinity.

SERMON ON THE MOUNT. X.

Verses 1-10.

Each child in class repeat one verse, then altogether.

What is this sermon called? Why? Where did the multitude come from? Why did they come so far? If somebody had done great thing for you, would you not go long way to hear and see him? And they found Jesus spoke as no one had ever spoken before—blessed words, all about goodness and happiness; not mere earthly happiness, but heavenly happiness lasting for ever. Now who are the happy ones.

(1) The poor in spirit—those who have no pride.

(2) They that mourn. Who will comfort them? Jesus. Better to mourn with Jesus than to be happy without Him.

(3) The meek—those who have the beautiful ornament of a meek and quiet spirit.

(4) Those who hunger and thirst after righteousness.

(5) The merciful—to enemies, to the weak, to animals.

(6) The pure in heart.

(7) The peacemakers.

(8) Those persecuted for righteousness' sake.

Who said these people would be happy? Is it true? Every word Jesus says always true. Is it the same now? Jesus same yesterday, to-day, and for ever—

always ready to comfort those who mourn, to fill those who want the Bread of Life and Living water; to make good the word of blessing given in this sermon so many years ago. You have heard the words, you have learnt them, and it is a very good thing if you know them well; but that is not all,—ask God to help you *do* them, learn to be meek and gentle like Jesus, to long after righteousness, to be merciful, pure and a peace-maker.

There was a lady once who had heard more than 7000 sermons, and though she had learned a great deal from many of them could only really remember two. You will most likely hear a good many sermons when you go to church, must listen, and try to learn, but do not forget this sermon, preached by Jesus Himself, the Sermon on the Mount.

Sixteenth Sunday after Trinity.

PRAYER.

"And all things, whatsoever ye shall ask in prayer, believing, ye shall receive."—*St. Matt.* xxi. 22.

Who said this? Look at verse 21, "*Jesus* answered and said." Who to? Disciples. What a beautiful promise for them. Was prayer new thing?—no; long ago, all over the world people prayed, felt they must—why? Why do you ask Mother for things? because you want them. Just the same with people—want things so badly, feel they must ask. What sort of things? success in what they are going to do, help in pain and sickness, for themselves or those they love, mercy from sins they know they are burdened with, freedom for fear of dying—all these things people want in all places, all over the world. If we want to receive, must ask right person—poor heathen ask idols, gods of wood and stone, for things; perhaps seen them in British Museum, ugly figures, no life, no goodness in them. Can they hear, answer, give? Chinese man prayed to the Sun for seven years. Could it hear and give? then to the Moon—no answer, very sad; heard of God, prayed to Him. Would God hear? "O Thou that hearest prayer, to Thee shall all flesh come" Ps. lxv. 2. Stand and repeat—Happy people that know the great God who answers prayer!

Can you remember some answers to prayer in Bible?

Abraham's servant Eliezer, Gen. xxiv. 15.	Nehemiah, Neh. ii. 4.
	Daniel, Dan. ix. 20.
Samuel, 1 Sam. vii. 9.	Jonah, Jonah ii.
Hezekiah, 2 Kings xix. 20.	David.

Can you tell me who made a prayer to Jesus on the cross? (Luke xxiii. 42). Could Jesus answer it? yes, Jesus is God. May children speak to God, pray to Him, ask for what they want? What did Jesus say about children? Do not always know how to pray; Holy Spirit will teach us; we must ask for that first—"Lord, teach us to pray." For very little ones this last text enough to found lesson upon, asking if they say prayers, and what.

Seventeenth Sunday after Trinity.

THE LORD'S PRAYER.

"Lord, teach us to pray."—*St. Luke* xi. 2.
"After this manner therefore pray ye: Our Father."—*St. Matt.*
vi. 9.

Verse we finished with last Sunday, began with to-day, is
a little prayer—Lord's Prayer is the answer. Why *Lord's*
Prayer? Who is our *Lord*? Do you know this prayer?
Who taught you? Where is it written? Woman
once lived to be very old, was taught Lord's Prayer
when little, never knew it was in the Bible, did not read
or care for Bible, did not think much about her prayer;
lady came one day, read the Lord's Prayer in Bible; old
woman so surprised, looked to see for herself, learned to
love Jesus Who had made such a beautiful prayer.
Disciples taught by Jesus Himself.

Lord's Prayer generally quite familiar to children, but
it is desirable if proper reverence can be maintained to
hear it repeated individually, to correct such mistakes as
" 'chart in heaven,"—" 'Kingdom come,"—" will b'done
earth, 'tis heaven "—sadly common, and fatal to proper
understanding of each word.

Explanations may follow if time allow.

" Our Father in heaven." Refer to Lesson 28.

" Hallowed be Thy name." Third Commandment, and
our duty to God, to " honour His holy Name and His

word." Happy if everybody did : cannot do this without His help. So pray that it may be done.

"Thy kingdom come." Happy, holy kingdom ! good to pray for.

"Thy will be done." God good, so His will good.

"As it is in heaven "—by the angels.

"Give us daily bread "—for body and soul.

"Forgive us "—merciful king.

"As we forgive "—unmerciful servant, must not be like him.

"Lead us not into temptation." Help us to follow Jesus our Leader.

"Deliver us from evil." See Lesson, Sermon on Mount vii.

What does *Amen* mean ?—So be it. Yes, we mean it. Told in the Bible all the people answered Amen. Hope you do in church and school. If you think of your prayer and ask Him in your heart God will send you a gracious answer, for Jesus' sake.

Eighteenth Sunday after Trinity.

BE THANKFUL.

"Be ye thankful."—*Col.* iii. 15.

Such a short verse, even the babies' class can learn it quite perfectly, only — words? three words. What does thank-ful mean—full of—? Almost the first thing children learn, if they have kind careful mothers, is to say " thank you " for what is given them.   Little babies told to say " ta " for crust or lump of sugar ; now bigger can say " thank you "—means " I thank you " for what you have given, or said, or done, for me.   Hope you don't forget.   Do you remember to thank mother for all the kind things she does—washing, dressing you, making clothes, getting dinner ready, so many things! and father for working hard to get money to buy food and pay the rent?   Give them loving thanks, let them see you love and thank them ever so much.   Now is there any one who loves even better than father or mother, does even more for us ?   God—God Who gives us parents, puts it into their hearts to be kind to their children, gives wind, rain, sunshine to make corn grow for bread, and so many other blessings (let children say as many as they can).   These all good for our bodies, but God takes care of souls too—food for them, kind teachers, Bible, and best of all " God so loved the world that He gave His Son," the Lord Jesus Christ, to be Saviour, Master, Leader, Living Bread to feed souls.   So precious! called " the unspeakable gift " 2 Cor. ix. 15—" Thanks be to God for His unspeakable gift " (repeat).   Then shall we not thank God too?

For little classes first text may be learned and repeated, lesson simply given, and teacher speak of saying grace. Mother gets dinner ready, but God makes corn to grow for bread ; even a little child can shut eyes, fold hands, and thank God while grace is said, or say for itself, " Thank

God for my good dinner." For bigger children point to
Duty to God in Catechism—" to give Him thanks "—
plain duty, do we do it? If you have asked anybody for
anything and they give it you, do you not thank them?
If you did not you would be very—? ungrateful. We
ask God to "give us this day our daily bread." He
gives it. Should we not thank Him? When we say grace
thank Him for the food for our bodies; when we
say prayers, thank Him for the best gift. What was
that?

Nineteenth Sunday after Trinity.

SPECIAL THANKSGIVING. HARVEST

(If there be harvest festival in church, this Lesson
may be used the same Sunday; or else some Sunday
chosen, end of August or early in September.)

"Thou preparest their corn, for so Thou providest for the earth."—
Psalm lxv. 10 (Prayer-book version).

What are the first words of this Psalm? "Thou, O
God—"; then it is "God" Who prepares the corn.
Verse 2, the very verse we learnt about prayer (Lesson
for 16th Sunday after Trinity). Has anybody prayed
about corn? Thousands, every day. What do they
say? "Give us this day our daily bread." Who
taught us to say that? Is the prayer answered?
Yes, even while they are praying God is preparing the

answer; getting ready the corn. How?—puts the
thoughts in men's hearts to sow the seed (if Teacher
can bring ear of corn it helps London children to understand), takes care of it, prepares it till it becomes wonderful waving harvest field. Let us turn to beautiful hymn
that tells us about it (S.P.C.K. Hymns, 282).

> "We plough the fields and scatter
> The good seed on the land,
> But it is fed and watered
> By God's Almighty Hand."

No better harvest lesson can be found than this hymn,
taken straight through, with comments and questions
leading children up through corn to verse 2, other
blessings of God in world around us; to special care for
His children, and our thankfulness for life, health, food,
all things bright and good, corresponding to "Creation,
preservation, and all the blessings of this life" in General
Thanksgiving, to which elder classes might refer, noticing
opportunity given for special thanksgiving after special
mercy.

How show our thankfulness? By gifts. God accepts
anything good we offer in love to Him—money, if ours to
give, kind words and deeds, trying to be good for His
sake. And best of all God loves to accept ourselves,
"our humble thankful hearts."

Twentieth Sunday after Trinity.

SPARROWS.

"Are not five sparrows sold for two farthings, and not one of them is forgotten before God."—*St. Luke* xii. 6.

(St. Matthew gives two for a farthing; St. Luke five for the double sum, one thrown in to complete the bargain as it were, and even that one of no monetary value, not forgotten before God.)

All London children know the sparrows, almost the only birds they do know—so bold and saucy, coming down to peck the horses' corn, such pretty little feathers and bright eyes; but so many of them, twittering, chirping, perching about everywhere. Good to eat? worth anything? very little: yet one great thing told about them—not forgotten before God. Wonderful thought! How can we believe it? Who says so? Read in Bible about it. Did you know there was anything about sparrow in Bible? Glad this verse is there, for it teaches us two things :—

(1) Nothing too small for God to mind; even the least of His creatures. He made them, knows all about them; and you see He does not forget them. This thought should help us to remember to be kind to all creatures—merciful (see Lesson 6 of those on Sermon on the Mount).

(2) And Jesus Himself teaches this lesson. Look

at verse 7, read it—Fear not ye therefore, hairs of your
head numbered, more value than many sparrows. How
comforting for disciples to hear; for us too, for we know
it is true of us too. If you had dear kind mother, who
cared for and looked after *little* things about the house,
would not she care for *big* things too. Cared for Pussy
and the pet bird, would she not care still more for *you*?
The loving good God cares for His little creatures, cares
still more for us—made us, keeps us, is our Father, gave
us Jesus Christ to die for us.

Twenty-first Sunday after Trinity.

FOXES.

"And Jesus said unto him, Foxes have holes, and birds of the air
have nests; but the Son of man hath not where to lay His head."—
St. Luke ix. 58.

Last Sunday had lesson from sparrows; do you think
of it sometimes as you see them hopping about? What
was the verse? What lesson did it teach?

Jesus taught lesson from others of His creatures beside
sparrows—hen, St. Matt. xxiii. 37; camel, St. Matt.
xix. 24; sheep, St. John x.; sheep and goats, St.
Matt. xxv.

Lessons to-day about foxes. Ever seen fox? one
in Zoological Gardens. For London children, if teacher
has picture or can make rough sketch, so much the

better. Country children hear and see them; they come to farmyard at night sometimes, steal nice fat hen, run away, very quick and sly—"Sly as a fox" people say. In winter gentlemen go out to hunt foxes —horses and dogs, red coats; fox runs away, everybody after him; one fox ran into cottage, jumped into the baby's cradle to hide! get right away sometimes, run home. What is their home? a *hole*—nice deep hole in ground, "foxes have holes." Who has taught them to make holes to live in? God. Even foxes, sly mischievous animals (place thought very sad and desolate when given up to foxes, Lam. v. 18), so mischievous that little *faults* compared to little foxes, yet they are not forgotten by God; He lets them make holes to live in.

If there be time the second half of text may be taken— birds in their nests, warm and comfortable—each little family in its own little home.

Now what is Jesus' lesson?—man had come, said he would follow Jesus, go where He went, live as He lived; Jesus wanted him to learn what hard life it was—no comfortable home, so said foxes have holes; even foxes; birds have nests; but "the Son of man (Jesus) hath not where to lay His head"—no home, no bed, walked from place to place, stayed with poor disciples, sometimes out all night—why? Did not God care for Him? Was He not God Himself? Yes, but Jesus became poor for our sakes, —"pleased not Himself," poor that we might be rich, no home on earth that we might have home in Heaven.

Twenty-second Sunday after Trinity.

LILIES OF THE FIELD.

"Consider the lilies of the field, how they grow; they toil not,
neither do they spin; and yet I say unto you that even Solomon in
all his glory was not arrayed like one of these."—*St. Matt.* vi. 28.

What was last Sunday's lesson about? What have foxes
got?—birds got? But Jesus had "no where"! What
lesson from sparrow? God remembers, and so still more
will care for—? Jesus teaches not only from animals—
foxes, sheep, etc., and birds, sparrows, hen; but even
beautiful lesson from flowers. Part of Sermon on the
Mount : people could see green grass before them, and
lovely wild flowers growing. Jesus knew it would be
easier for them to learn from something they could see—
said, "Consider the lilies," look at, notice them (nice if
Teacher could bring flower to show its beauty, colour more
beautiful, texture more fine soft and delicate than any silk
or satin), see how they grow—do not work to get money,
do not spin to make cloth to wear; yet dressed so beauti-
fully, even Solomon not so grand as they. Who was
Solomon? (refer to 1 Kings x. or 2 Chron. ix). Describe
glory of Solomon, gold, fine linen, purple stuffs—Queen of
Sheba came long journey to see, was astonished at his
glory) and yet Jesus says these little flowers, these lilies
of the field—dressed better than he. Now what is the

lesson Jesus teaches from this? what from the sparrow? —that God remembers and feeds sparrow, so will much more remember and feed us His children. Same lesson now about our clothes. Read beginning of verse 28, "take no thought"—be not anxious (see Revised Version) about your raiment (clothes); God clothes the flowers, will clothe you; verse 33, if we "seek first the kingdom of God and His righteousness," these things— food and clothes, daily bread—given to us.

Not only for bodies, but souls—"My soul shall be joyful in my God, for He hath clothed me with the garments of salvation, He hath covered me with robe of righteousness," Isa. lxi. 10; "clothed with humility," 1 Pet. v. 5; "with the ornament of a meek and quiet spirit." Not only now but by and by—white robes of Heaven.

Twenty-fifth Sunday after Trinity.

GATHER UP THE FRAGMENTS.

"Gather up the fragments that remain, that nothing be lost."— *St. John* vi. 12.

Jesus by the sea-side again: great many people followed Him, often did. Tell me some times—when Jairus came, and Sermon on the Mount, and seeing the—? multitudes—same word now, what does it mean? why had they come? because they had seen the miracles

(what are miracles?) that Jesus did for the sick.   Let
Teacher describe scene—little grassy hill, Jesus sitting,
with disciples round Him, looking at the crowds below,
felt sorry for them, some were sick (St. Matt. xiv. 14),
all knew what sorrow was, all had sinned, were like lost
sheep and did not know Jesus was their Shepherd (St.
Mark vi), began to teach and heal them, went on all day
till evening; people so eager to listen, did not think of
being tired or going away; but Jesus knew, been all day
without food, were—? hungry: 5000 people, no shops to
buy bread; could you have fed them?   Could Jesus?
We shall see.   First asked if Philip could tell any way.
No—cost great deal of money, and then no shop.   Tell
about boy who had brought five loaves, two little fishes,
enough for twenty men, but there were 5000!   Now
comes the wonderful part, the miracle!   Jesus took
them and *made them enough.*   Describe the people sitting
down in rows on the grass (like school treat perhaps),
Jesus giving thanks, disciples giving out food received
from Him.   At last all had eaten, all had had enough—
how do we know there was really enough?   Two ways;
first told so, "as much as they wanted," second, there
was some over.   Poor family sharing scanty dinner, eat
up every crumb, only have some left when there is
plenty.   Did Jesus care about what was left?   Yes,
gave directions to disciples about the bits—words of our
text told them what to do—(repeat).   We learnt that
God does not forget sparrows, see now that He even

remembers little bits of bread.   How careful we should be not to waste.   And while the miracle helps us, as it did the 5000 men, to feel the greatness of Jesus, let it help us to think of the little fragments of food, time, opportunities, teaching—had much given, what can you gather up? These made twelve baskets, what will yours make?

Printed at the University Press, Oxford

By Horace Hart, *Printer to the University*

Society for Promoting Christian Knowledge.

Publications on
THE CHRISTIAN EVIDENCE.

BOOKS.

Price.

Natural Theology of Natural Beauty (The).
By the Rev. R. St. John Tyrwhitt, M.A. Post 8vo.
Cloth boards 1 6

Steps to Faith.
Addresses on some points in the Controversy with Unbelief.
By the Rev. Brownlow Maitland, M.A. Post 8vo.
Cloth boards 1 6

Scepticism and Faith.
By the Rev. Brownlow Maitland. Post 8vo. *Cloth boards* 1 4

Theism or Agnosticism.
An Essay on the grounds of Belief in God. By the Rev.
Brownlow Maitland, M.A. Post 8vo.............*Cloth boards* 1 6

Argument from Prophecy (The).
By the Rev. Brownlow Maitland, M.A., Author of
"Scepticism and Faith," &c. Post 8vo. ....*Cloth boards* 1 6

Being of God, Six Addresses on the.
By C. J. Ellicott, D.D., Bishop of Gloucester and Bristol.
Small Post 8vo. ....*Cloth boards* 1 6

Modern Unbelief: its Principles and Charac-
TERISTICS. By the Right Rev. the Lord Bishop of Gloucester
and Bristol. Post 8vo. ....*Cloth boards* 1 6

Some Modern Religious Difficulties.
Six Sermons preached, by the request of the Christian
Evidence Society, at St. James's, Piccadilly, on Sunday
Afternoons after Easter, 1876 ; with a Preface by his Grace
the late Archbishop of Canterbury. Post 8vo. *Cloth boards* 1 6

Some Witnesses for the Faith.
Six Sermons preached, by the request of the Christian
Evidence Society, at St. Stephen's Church, South Kensing-
ton, on Sunday Afternoons after Easter, 1877. Post 8vo.
Cloth boards 1 4

Theism and Christianity.
Six Sermons preached, by the request of the Christian
Evidence Society, at St. James's, Piccadilly, on Sunday
Afternoons after Easter, 1878. Post 8vo.......*Cloth boards* 1 6

Price.

When was the Pentateuch Written?

By George Warington, B.A., Author of "Can we Believe in Miracles?" &c. Post 8vo....................... *Cloth boards* 1 6

The Analogy of Religion.

Dialogues founded upon Butler's "Analogy of Religion." By the late Rev. H. R. Huckin, D.D., Head Master of Repton School. Post 8vo. .... *Cloth boards* 3 0

"Miracles."

By the Rev. E. A. Litton, M.A., Examining Chaplain of the Bishop of Durham. Crown 8vo. .... *Cloth boards* 1 6

Moral Difficulties connected with the Bible.

Being the Boyle Lectures for 1871. By the Ven. Archdeacon Hessey, D.C.L. Preacher to the Hon. Society of Gray's Inn, &c. FIRST SERIES. Post 8vo. ...*Cloth boards* 1 6

Moral Difficulties connected with the Bible.

Being the Boyle Lectures for 1872. By the Ven. Archdeacon Hessey, D.C.L. SECOND SERIES. Post 8vo.
Cloth boards 2 6

Prayer and recent Difficulties about it.

The Boyle Lectures for 1873, being the THIRD SERIES of "Moral Difficulties connected with the Bible." By the Ven. Archdeacon Hessey, D.C.L. Post 8vo.
Cloth boards 2 6

The above Three Series in a volume ....*Cloth boards* 6 0

Historical Illustrations of the Old Testament.

By the Rev. G. Rawlinson, M.A., Camden Professor of Ancient History, Oxford. Post 8vo ....*Cloth boards* 1 6

Can we Believe in Miracles?

By G. Warington, B.A., of Caius College, Cambridge. Post 8vo... *Cloth boards* 1 6

The Moral Teaching of the New Testament

VIEWED AS EVIDENTIAL TO ITS HISTORICAL TRUTH. By the Rev. C. A. Row, M.A. Post 8vo....................*Cloth boards* 1 9

Scripture Doctrine of Creation.

By the Rev. T. R. Birks, M.A., Professor of Moral Philosophy at Cambridge. Post 8vo..............................*Cloth boards* 1 6

Price.
s. d.

The Witness of the Heart to Christ.

Being the Hulsean Lectures for 1878. By the Rev. W. Boyd
Carpenter, M.A. Post 8vo. ....*Cloth boards* 1 6

Thoughts on the First Principles of the Positive

PHILOSOPHY, CONSIDERED IN RELATION TO THE HUMAN
MIND. By the late Benjamin Shaw, M.A., late Fellow
of Trinity College, Camb. Post 8vo. ....*Limp cloth* 0 8

Thoughts on the Bible.

By the late Rev. W. Gresley, M.A., Prebendary of Lichfield.
Post 8vo. ....*Cloth boards* 1 6

The Reasonableness of Prayer.

By the Rev. P. Onslow, M.A. Post 8vo. ....*Paper cover* 0 8

Paley's Evidences of Christianity.

A New Edition, with Notes, Appendix, and Preface. By
the Rev. E. A. Litton, M.A. Post 8vo..........*Cloth boards* 4 0

Paley's Natural Theology.

Revised to harmonize with Modern Science. By Mr. F. le
Gros Clark, F.R.S., President of the Royal College of
Surgeons of England, &c. Post 8vo. ....*Cloth boards* 4 0

Paley's Horæ Paulinæ.

A new Edition, with Notes, Appendix, and Preface. By
J. S. Howson, D.D., Dean of Chester. Post 8vo. *Cloth boards* 3 0

Religion and Morality.

By the Rev. Richard T. Smith, B.D., Canon of St. Patrick's,
Dublin. Post 8vo. ....*Cloth boards* 1 6

The Story of Creation as told by Theology

AND SCIENCE. By the Rev. T. S. Ackland, M.A. Post 8vo.
Cloth boards 1 6

Man's Accountableness for his Religious Belief.

A Lecture delivered at the Hall of Science, on Tuesday,
April 2nd, 1872. By the Rev. Daniel Moore, M.A., Holy
Trinity, Paddington. Post 8vo. ....*Paper cover* 0 3

The Theory of Prayer; with Special Reference

TO MODERN THOUGHT. By the Rev. W. H. Karslake,
M.A., Assistant Preacher at Lincoln's Inn, Vicar of
Westcott, Dorking. Post 8vo. ....*Limp cloth* 1 0

The Credibility of Mysteries.

A Lecture delivered at St. George's Hall, Langham Place.
By the Rev. Daniel Moore, M.A. Post 8vo......*Paper cover* 0 3

Price.
s. d.

The Gospels of the New Testament: their
GENUINENESS AND AUTHORITY. By the Rev. R. J. Crosthwaite, M.A. Post 8vo...............................*Paper cover* 0 3

Analogy of Religion, Natural and Revealed,
TO THE CONSTITUTION AND COURSE OF NATURE: to which are added, Two Brief Dissertations. By Bishop Butler. NEW EDITION. Post 8vo.............................*Cloth boards* 2 6

Christian Evidences:
intended chiefly for the young. By the Most Reverend Richard Whately, D.D. 12mo.. .... *Paper cover* 0 4

The Efficacy of Prayer.
By the Rev. W. H. Karslake, M.A., Assistant Preacher at Lincoln's Inn, &c. &c. Post 8vo. .... *Limp cloth* 0 6

Science and the Bible: a Lecture by the Right
Rev. Bishop Perry, D.D. 18mo. *Paper cover* 4d., or *Limp cloth* 0 6

A Lecture on the Bible. By the Very Rev.
E. M. Goulburn, D.D., Dean of Norwich. 18mo. *Paper cover* 0 2

The Bible: Its Evidences, Characteristics, and
EFFECTS. A Lecture by the Right Rev. Bishop Perry, D.D. 18mo...*Paper cover* 0 4

The Origin of the World according to
REVELATION AND SCIENCE. A Lecture by Harvey Goodwin, M.A., Bishop of Carlisle. Post 8vo....*Cloth boards* 0 4

On the Origin of the Laws of Nature.
By Sir Edmund Beckett, Bart. Post 8vo.......*Cloth boards* 1 6

What is Natural Theology?
Being the Boyle Lectures for 1876. By the Rev. Alfred Barry, D.D., Bishop of Sydney. Post 8vo.......*Cloth boards* 2 6

* *For List of TRACTS on the Christian Evidences, see the Society's Catalogue B.*

LONDON:
SOCIETY FOR PROMOTING CHRISTIAN KNOWLEDGE,
NORTHUMBERLAND AVENUE, CHARING CROSS, W.C.;
43, QUEEN VICTORIA STREET, E.C.;
26, ST. GEORGE'S PLACE, HYDE PARK CORNER, S.W.
BRIGHTON: 135, NORTH STREET.

CPSIA information can be obtained at www.ICGtesting.com
Printed in the USA
BVOW08s2351160315

391994BV00016B/126/P